Statistics

Statistics for Geographers

by

Martin Mowforth M.Sc.

Formerly Head of Geography
Sladebrook High School
London

Nelson

Thomas Nelson and Sons Ltd
Nelson House   Mayfield Road
Walton-on-Thames   Surrey
KT12 5PL   UK

51 York Place
Edinburgh
EH1 3JD   UK

Thomas Nelson (Hong Kong) Ltd
Toppan Building 10/F
22A Westlands Road
Quarry Bay   Hong Kong

Thomas Nelson Australia
102 Dodds Street
South Melbourne   Victoria 3205
Australia

Nelson Canada
1120 Birchmount Road
Scarborough   Ontario
M1K 5G4   Canada

First published by Harrap Limited 1979
ISBN 0-245-53399-0

This edition published by Thomas Nelson and Sons Ltd 1985
ISBN 0-17-444241-6
NPN 9 8 7

Printed in Hong Kong

Preface

A large part of the material for this book was prepared for a statistics course designed for teachers of the new London University A level geography syllabus. It became clear to the author, a practising teacher, during and after that course, that most of the statistics texts are aimed at the 'top-flight' A level student and first-year undergraduate level. A long search is required to find texts, not just suitable for, but intended directly for, the average A level geography student. The material prepared for that course, then, has been adapted specifically for such a student, and this book is the result of that adaptation.

Acknowledgements

For permission to reproduce material, I acknowledge the following: the Longman Group Ltd for Figs. 15, 20 and 22, which are taken from Fig. 33, p. 153, Fig. 28, p. 120 and Fig. 34, p. 188, respectively, of *Statistical Methods and the Geographer*, by S. Gregory, fourth edition, 1978. Also for Figs. 15 and 20 I thank the Biometrika Trustees—these figures are based on Tables 8 and 12, respectively, from *Biometrika Tables for Statisticians*, Vol. 1, third edition, 1966. Fig. 30 is reproduced here with the permission of the Institute of British Geographers, publishers of *Area*, Vol. 8, No. 3, from which it was taken. For the same illustration I thank David Ebdon for his co-operation and notes. On the subject of nearest neighbour analysis I owe much to Dr D. A. Pinder, especially for up-dating my knowledge of the technique. I regret that space did not allow the inclusion of full details of Dr Pinder's work as it appears in *Area*, Vol. 10, No. 5, 'Correcting underestimation in nearest neighbour analysis'.

For Table 33 and much of the information in chapter 20 I thank Jim Rose of Birkbeck College. Also of Birkbeck College, Dr P. W. Lewis deserves much credit and gratitude for being the provider of so much food for thought and for many of the ideas discussed throughout the book. A number of ideas included here are done so only after discussion with Paul Cope, to whom I am indebted for advice and a worked example. Many hours have been spent editing the original manuscript; for this, Ian Hyde's efforts, help and tolerance are much appreciated.

Acknowledgements for ideas, advice, encouragement and data supply, or some combination of these four, are also deserved by Bruce Grierson, June Mowforth and the Statistics Division of the Department of Employment.

All errors and omissions, of course, remain the sole responsibility of the author.

Contents

Figures

Tables

Introduction

Recently the study of geography has dropped the idea of uniqueness that led to lengthy and detailed regional descriptions as a major part of the art of geography. The treatment of the subject as an encyclopedic discipline has declined, although it still lingers in some parts. The Advanced Level of the subject particularly suffered from this malaise, and past sixth-form students will remember that in the consideration of their choice of subjects they faced the realization that geography was not especially difficult in its conceptual demands, but consisted of a very great amount of work and much dependence on the memory of facts.

Whilst it is still accepted that all objects are unique, it is now widely agreed that this fact does not preclude similarities between objects. Two chairs of the same design are clearly unique in that chair A is Chair A and chair B is chair B, and at any one time neither of them can be displaced by the other. These same two chairs, however, have similarities and from the observation of these similarities certain generalizations can be made—the chairs are brown and have four legs, etc. From patterns of and collections of similarities, then, generalizations and theories, and possibly even more stringent laws, can be put forward.

It is tempting to geographers, as they are not statisticians, to disregard or avoid a vital step in the formulation of these theories: that is, the careful and thoughtful analysis of their observations before the generalization is made. Too often the 'look-and-say' method is used, where one observes an example of a particular phenomenon and dives straight into the theory or accepts too readily the theory or assertion of another.

Only with something a little more rigorous than the 'look-and-say' method can geographers have confidence in the theories and ideas with which they deal. The data or observations from which their conclusions are drawn should be subject to close scrutiny, and the theory should fit the facts if it is going to carry any weight. This is where an understanding of some basic statistical techniques is essential to the geographer.

The new Advanced Level geography syllabuses place far more emphasis on the development of conceptual understanding than on the accumulation of factual knowledge. If the concepts and theories of the new geography are to be understood it is important that the steps taken in reaching them should be followed by the student. As many geographic observations are collected and arranged in tables of numerical data, statistical analysis is a very necessary step on the way to the formulation, development and understanding of theories.

It is the attempt of this book to cover the statistics demanded by these new Advanced Level geography syllabuses in a simple and uncluttered way. Further, it is seen as vital that the assumptions behind and limitations of each technique should be clearly outlined in this, probably the student's first meeting with the use of statistics in practical situations; these assumptions and limitations are stressed in this book.

1 The data

Ways of manipulating statistical data are in large part controlled by the nature of the data themselves. It is therefore necessary to begin with an outline of some of the basic characteristics of all types of statistical data.

Populations. Our data is contained in populations. Any population is the total number of objects under consideration. All objects in the population are identified by one attribute that they have in common. For example, we might consider the human population of Great Britain, in which case we would use census data for analysis. We would therefore have access to the full population—i.e., *N*.

Samples. More often, however, we have to use a sample of the full population—i.e. *n*, as sets of data containing the full population are uncommon in geography. A sample is simply a portion of the full population.

In order to prevent bias in the sample, and if it is to be a worthwhile and meaningful representation of the full population, it should be made by a random procedure, in which case it is a *random sample.* To help the geographer/researcher a table of random numbers can be used. These can be found in *Cambridge Elementary Statistical Tables* by Lindley and Miller (C.U.P.), or in numerous other recent works of geography. Assuming that our study requires a sample of points over an area, a grid system must first be imposed upon a map of the area. The table of random numbers can then be employed to generate grid references at random. The pattern of sample points so produced is shown in Fig. 1(a). Alternatively, for other purposes the table can simply be used to generate numbers at random.

Other sampling procedures are also acceptable. *Systematic sampling* would produce a series of points that are equally spaced over the area under study, as shown in Fig. 1(b). Of course some decisions are left with the sampler, and pre-judgement can still therefore occur as there are numerous different ways of taking a systematic sample. A square or triangular pattern can be chosen; the distance between the points and the number of points must be selected carefully; and the origin of the system of points on a grid square must be determined.

Fig. 1 Three sampling patterns

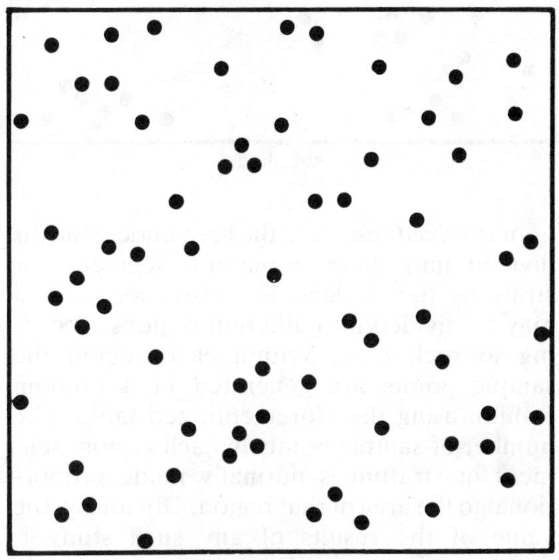

(a) random

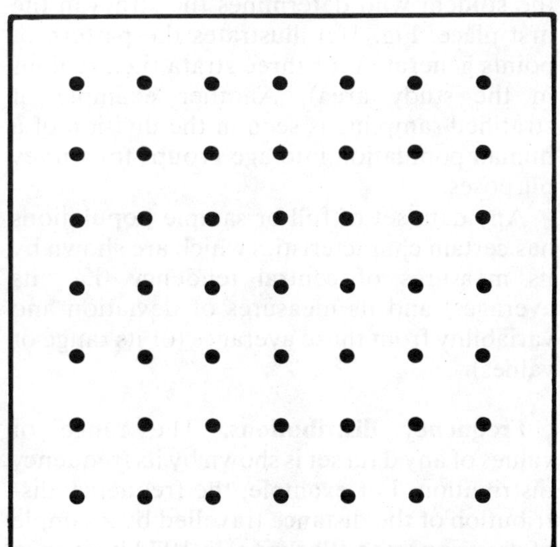

(b) systematic

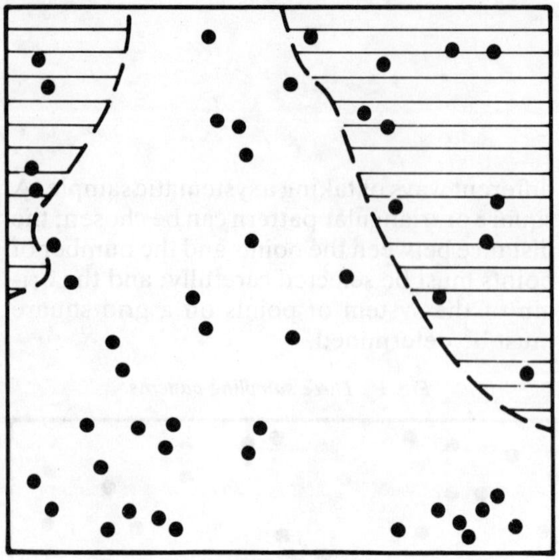

(c) stratified

In *stratified sampling* the area under study is divided into different natural segments or strata by the student. For instance, an area may be divided into different regions according to rock type. Within each region the sample points are generated in a random fashion using the aforementioned table. The number of sample points in each region, segment or stratum is normally made proportional to the area of that region. Obviously, the value of the results of any such study is dependent to some extent on the judgement of the student who determines the strata in the first place. Fig. 1(c) illustrates the pattern of points generated for three strata (i.e., regions in the study area). Another example of stratified sampling is seen in the division of a human population into age groups for survey purposes.

Any data set of full or sample populations has certain characteristics which are shown by its measures of central tendency (i.e., its averages) and its measures of deviation and variability from those averages (or its range of values).

Frequency distributions. The range of values of any data set is shown by its frequency distribution. For example, the frequency distribution of the distance travelled by a sample of day trippers to Blackpool in 1974 is given in Fig. 2. This frequency distribution is shown as

Fig. 2 Frequency distribution of the distance travelled by day trippers to Blackpool

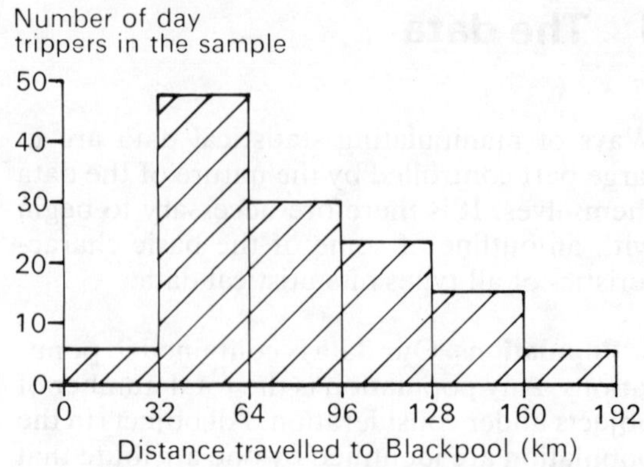

a bar graph or histogram, but from this a line can be interpolated (see chapter 18) to give a frequency curve. This is obtained by plotting the mid-interval points, as shown in Fig. 3.

Fig. 3 Frequency curve of the distance travelled by day trippers to Blackpool

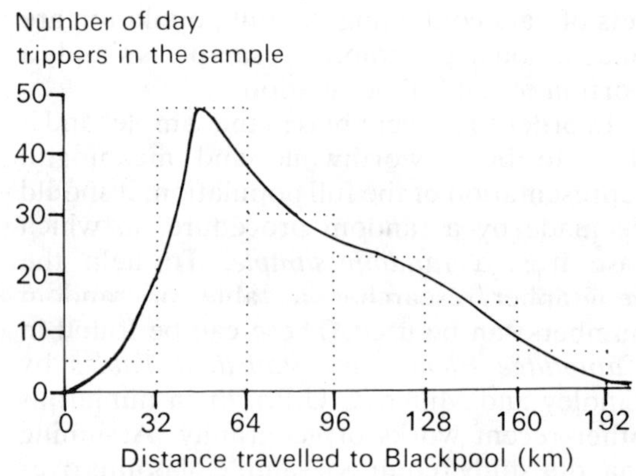

The normal (frequency) distribution is a distribution whose measures of central tendency (the arithmetic mean, the median and the mode) are equal to each other, and which has a frequency curve symmetrical on either side of its averages (see Fig. 4). Other properties of the normal distribution are given in chapter 5. This is the normal frequency curve *not* in that it is the most commonly found, but in that it represents an ideal or standard distribution against which others can be compared. It is found in data sets of natural phenomena or in data sets of the occurrences of any event in which chance plays a full share.

*Fig. 4** *The frequency curve of the normal distribution*

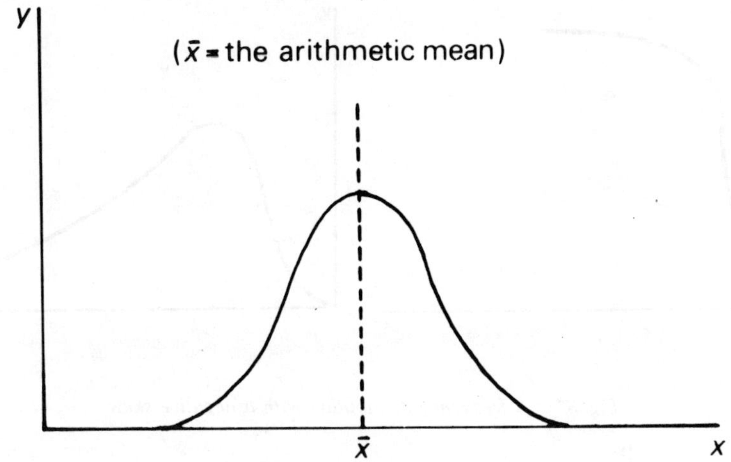

Table 1 *Cumulative distance travelled by day trippers to Blackpool in 1974*

Distance travelled (km)	*Cumulative number of day trippers in sample*
0–32	6
33–64	54
65–96	84
97–128	107
129–160	122
161–192	127

A cumulative frequency curve, or ogive, can also be shown to represent a data set. Fig. 5 shows the cumulative frequency curve for the data set already used (Table 1). It is obtained by adding the total for each frequency to the total of the previous frequency.

The normal ogive is obtained from a symmetrical frequency distribution, and is shown in Fig. 6.

However, the frequency curve of any distribution may be unbalanced. This is skewness, and can be either positive or negative, as illustrated in Figs. 7 and 8.

* In Figs 4, 6, 7 and 8 the x axis shows the range of values in the sample or population (e.g., the distance travelled by day trippers to Blackpool in the example used here) and the y axis represents the number of values in the sample or population.

Fig. 5 *The cumulative frequency curve of the distance travelled by day trippers to Blackpool*

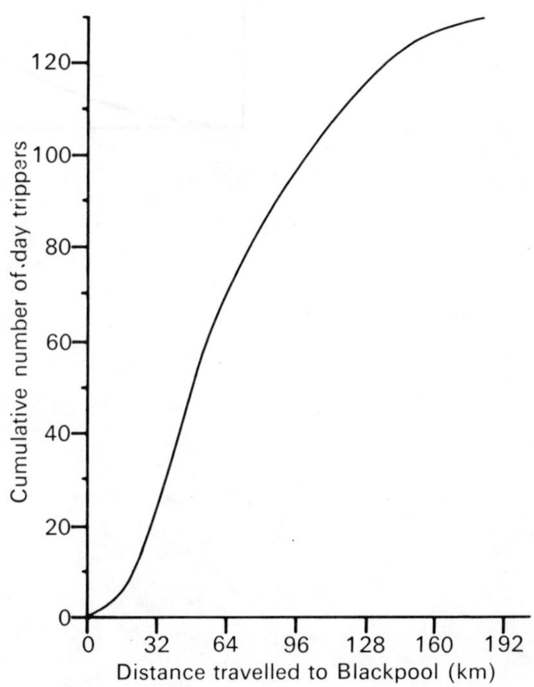

Any frequency distribution can be described by its measures of central tendency and by its measures of deviation. These are detailed in the following chapters.

Fig. 6   The normal cumulative frequency distribution*

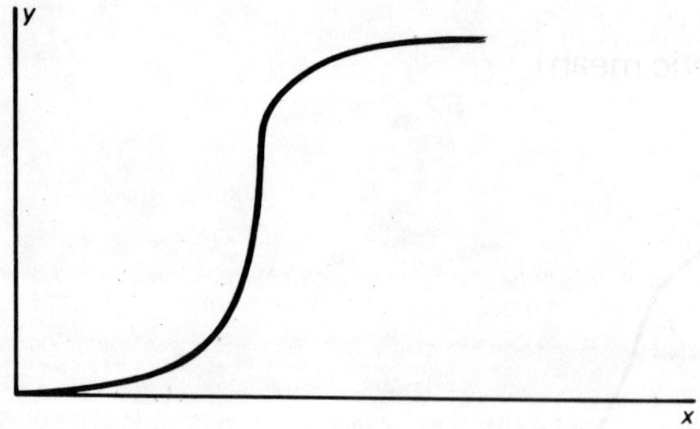

Fig. 7   A frequency distribution with a positive skew*

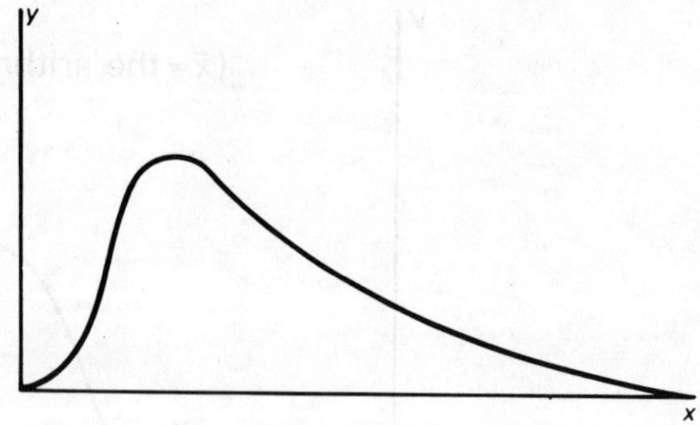

Fig. 8   A frequency distribution with a negative skew*

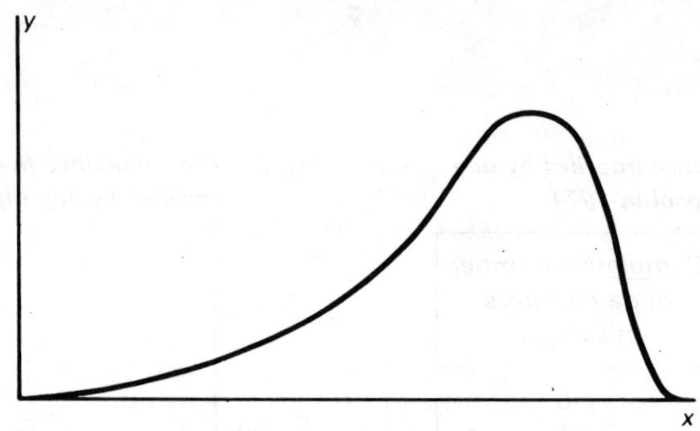

2 Measures of central tendency

All of these may be called averages, and they include the arithmetic mean, the median, and the mode.

The arithmetic mean. Values in a data set tend to cluster around the centre of the normal distribution. One of the best ways of illustrating and describing the cluster of values is the arithmetic mean—that value that is numerically most representative of the data set.

It is also the most widely used measure of central tendency and is easy to calculate. It is denoted as $\bar{x}$ or μ (Greek letter mu) and is found by using the formula:

$$\bar{x} = \frac{\Sigma x}{n}$$

where:

Σ = the sum of (Greek letter capital sigma)

Table 2 Protein intake in selected Third World countries, 1971

Country	*Grams of protein per person per day*
Argentina	88
Bolivia	51
Brazil	66
Ecuador	52
Mexico	66
Peru	54
Venezuela	66
Ethiopia	69
Gambia	37
Ghana	49
Kenya	64
Morocco	54
Nigeria	52
South Africa	80
Sudan	64
Tanzania	58
Uganda	50
UAR	81
Jamaica	53
Iran	50

(Source: 'Statistics of Food' 1971)

x = the values in the data set
n = the total number of values in the data set

Applying this to the data set shown in Table 2, we see that:

$$n = 20$$
$$\Sigma x = 1{,}204$$

and therefore that:

$$\bar{x} = \frac{1{,}204}{20}$$
$$= 60{\cdot}2 \textit{ grams}$$

This is the most accurate measure of central tendency, but it suffers from the disadvantages of being, firstly, affected by extreme values, and secondly, unreliable when the data set contains only a few values.

The median. If we arranged our data set in rank order, from highest to lowest, as in Table 3, the median is that figure which is the mid-point in the list. In our data set, then, the median falls at the mid-point between the tenth and eleventh values, i.e., between 58 and 54 (marked 'Md' in Table 3.) The median is therefore 56 grams.

Had the tenth and eleventh values been 55 and 54, then the median value would have been 54·5. This is quite acceptable despite the decimal point. Had there been 21 values in the data set, then the median would have been the eleventh value in the list.

A disadvantage of this measure is that equal weight is given to each item in the list regardless of the value of that item. This may be an advantage in some cases, but it is generally unwise to ignore the value of individual items in the data set.

The mode. As the name suggests, this is the most frequently occurring value in the data set. So, in the data set shown in Table 3 the mode is 66. A data set is not necessarily unimodal, however. For instance, had there been another value of 54 in the above data set it would have been bimodal, as there would have

Table 3 Rank order of Third World countries according to their protein intake, 1971

Country	Grams of protein per person per day	
Argentina	88	
UAR	81	
South Africa	80	
Ethiopia	69	
Venezuela	66	← UQ
Mexico	66	
Brazil	66	
Sudan	64	
Kenya	64	
Tanzania	58	← Md
Peru	54	
Morocco	54	
Jamaica	53	
Nigeria	52	
Ecuador	52	← LQ
Bolivia	51	
Uganda	50	
Iran	50	
Ghana	49	
Gambia	37	

UQ = Upper quartile
Md = Median
LQ = Lower quartile
See chapter 5.

been two values that occurred on three occasions each.

In the histogram the modal column is the highest one—i.e., the most frequently occurring class of values. In Fig. 2 this modal class value is the class 32 to 64 km from Blackpool. On a frequency curve the modal value is shown by the peak of the curve.

Grouped data. On frequent occasions and in numerous situations the table of original data may not be available to the geographer. It may be in the form of groups or classes, as is the protein intake data in Table 4. If this is the case it is still possible to calculate an approximate mean for the data set.

The second column of Table 4 gives the mid-point of each class and the third indicates how many of the 20 values fall in each class. The product of the items in these two columns (fx_0) is given in the final column and these products are then totalled to give Σfx_0. The simple formula for the approximate mean is:

$$\bar{x} = \frac{\Sigma fx_0}{n} = \frac{\Sigma fx_0}{\Sigma f}$$

Table 4 Protein intake in selected Third World countries, 1971, arranged in groups

Class (grams of protein)	Mid-point value x_0	Frequency f	(fx_0)
30–39	34·5	1	34·5
40–49	44·5	1	44·5
50–59	54·5	9	490·5
60–69	64·5	6	387·0
70–79	74·5	0	0
80–89	84·5	3	253·5
		$\Sigma f = 20$	$\Sigma fx_0 = 1{,}210$

where:

$\bar{x}$ is the approximate mean
f is the number of items in each class
x_0 is the mid-point of each class
n is the total number of values in the data set.

In this example:

$$\bar{x} = \frac{1{,}210{\cdot}0}{20}$$

$$\therefore \bar{x} = 60{\cdot}5 \text{ grams}$$

The difference between this result and the actual arithmetic mean is only 0·3 grams.

This statistical technique may be necessary when the only available data is grouped, but it may also be more convenient when the data set contains a very large number of items and a calculator cannot be found. In other words, it may be easier to group the data and calculate the grouped mean rather than go through the lengthy process of summing all the individual values. It is essential to bear in mind, though, that the result is only an approximation to the arithmetic mean.

From grouped data it is also possible to estimate the median, and this is best done by plotting the ogive of the data set. Fig. 9 shows the cumulative frequency distribution in percentage form plotted from the protein intake data given in Table 4. Obviously the median value occurs half way along the y axis of cumulative frequency—i.e., at the 50% mark. The dashed line on Fig. 9 shows the estimated median to occur at 59 grams, and this can be compared with the actual median of 56. This discrepancy arises because the estimated median does not account for the individual values, but simply for the groups or classes of values. Like the estimated mean, the most useful application of the estimated median is seen when the original data, from which the groups are formed, are not available.

Fig. 9 Cumulative frequency curve for protein intake data

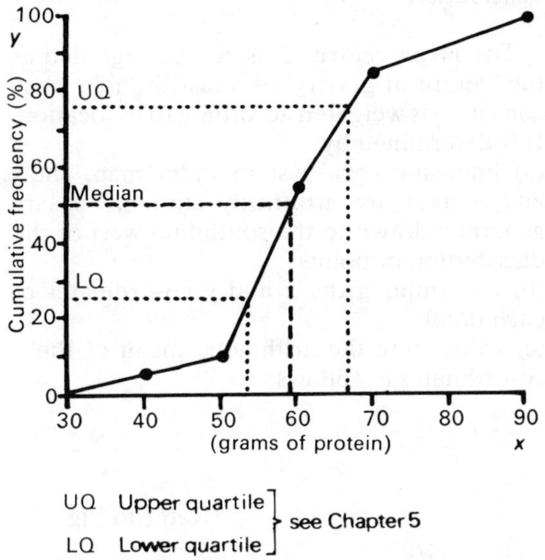

So, the measures of central tendency for the data set shown in Tables 2 and 3 are as follows:

arithmetic mean	60·2 grams
grouped mean	60·5 grams
mode	66 grams
modal class	50–59 grams
median	56 grams
estimated median	59 grams

If the frequency distribution of the data set is normal—i.e., ideal and symmetrical—these measures would all have the same value.

If, however, the frequency distribution is asymmetrical with a negative skew (Fig. 8) then the mode is greater than the median, which is greater than the arithmetic mean. Expressed mathematically, this is:

negative skew: mode > median > $\bar{x}$*

If the distribution displays a positive skew, then the arithmetic mean is greater than the median, which is greater than the mode—mathematically:

positive skew: $\bar{x}$ > median > mode

Using any one of these measures of central tendency on its own to represent a data set is a little unreliable, particularly with a set of data that has a wide frequency distribution—i.e., its extreme values far apart from each other. If one is searching purely for a measure of central tendency, then it would probably be wiser to use all three in conjunction with each other.

Also these measures do not yield any information regarding the scatter of values around a central point or value, and, without this scatter, analysis of any data set is incomplete. There are several methods used to describe this scatter and most of them employ the notion of deviation from the arithmetic mean. These are discussed in chapter 5.

* > This notation represents 'is greater than'.

3 Measures of central tendency applied to areal distributions

Measures of central tendency can not only be applied to sets of numbers, but can also be used to determine the central point of a distribution of points or locations over space. This is of particular importance to geographers, of course, as the results are expressed as locations on a map rather than by numbers.

For both mean and median centres of a distribution of points a grid system is imposed upon the map. In this way each location can be given a grid reference, or co-ordinates for the x and y axes. Fig. 10 shows the distribution of the suppliers of raw materials to a factory manufacturing pens. It is possible to calculate the mean and median centres of this point distribution.

Fig. 10 The mean and median centres of a point distribution

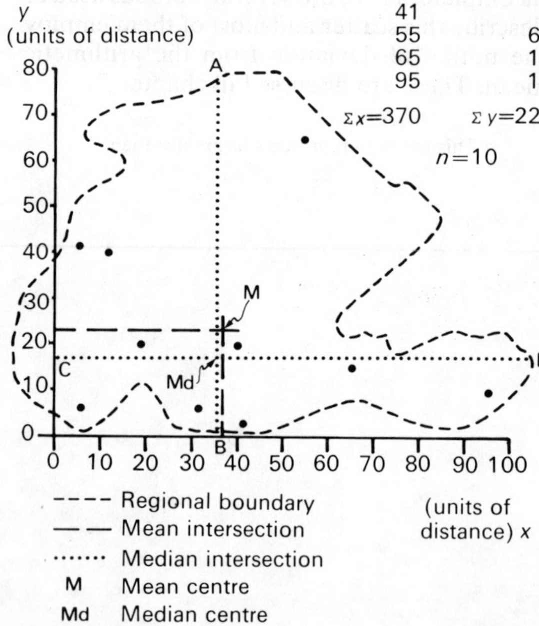

x axis	y axis
6	7
6	42
12	40
19	21
31	6
40	20
41	3
55	65
65	15
95	10
$\Sigma x=370$	$\Sigma y=229$
	$n=10$

The mean centre. This can be regarded as the 'centre of gravity' of a distribution, since each item is weighted according to its location. It is determined by:

(a) imposing a grid system on the map. The x and y axes are arbitrarily chosen, but are generally drawn to the south and west of the distribution of points;

(b) determining the x and y co-ordinates of each point;

(c) calculating the arithmetic mean of the x co-ordinates as follows:

$$\frac{6 + 6 + 12 + 19 + 31 + 40 + 41 + 55 + 65 + 95}{10}$$

(Refer to Fig. 10)

$$= \frac{370}{10} = 37{\cdot}0$$

(d) calculating the arithmetic mean of the y co-ordinates as follows:

$$\frac{3 + 6 + 7 + 10 + 15 + 20 + 21 + 40 + 42 + 65}{10}$$

$$= \frac{229}{10} = 22{\cdot}9$$

(e) plotting these co-ordinates to show the mean centre (M).

The median centre. This is the point of intersection of two perpendicular lines (usually drawn from south to north and from west to east,) each of which bisects the distribution of points. Thus, on Fig. 10, line AB divides the distribution along the x axis, and line CD along the y axis. The intersection of these lines gives the median centre (Md).

The median centre is a measure of the minimum aggregate distance (to the centre) and is therefore a theoretical optimum location for service centres.

Uses and applications. These techniques can be used to determine:
(a) the central location of any point distribution, such as shops in a town, finance institutions in a city, perceptions of a town centre, etc.
(b) the difference between an actual location and the theoretical optimum location.

Previous geographical works have used these techniques to show such trends as:
(a) an eastwards shift in the population of the USSR over a number of decades;
(b) a shift in the centre of gravity of the production of various agricultural crops in the USSR;
(c) a shift of orange-growing progressively southwards in Florida.

Problems. Two major problems arise, and both of these should be fully understood by a prospective employer of these techniques:
(a) The grid system is superimposed upon the point distribution in an arbitrary way. Although a north–south and east–west grid is the most common, it is not necessary for all grids to conform to this; but if the grid system were orientated north-west/south-east the median centre could alter its location. A change in the orientation should not, however, affect the location of the mean centre.
(b) The common geographical problem of map scale and the size of the area under study can also have a bearing upon the use of these techniques. If the size of the area under consideration is altered, although the mean and median centres may not change locations, the image of the point distribution could be very different. Fig. 11 illustrates this problem.

Fig. 11 Changes in area size and impression

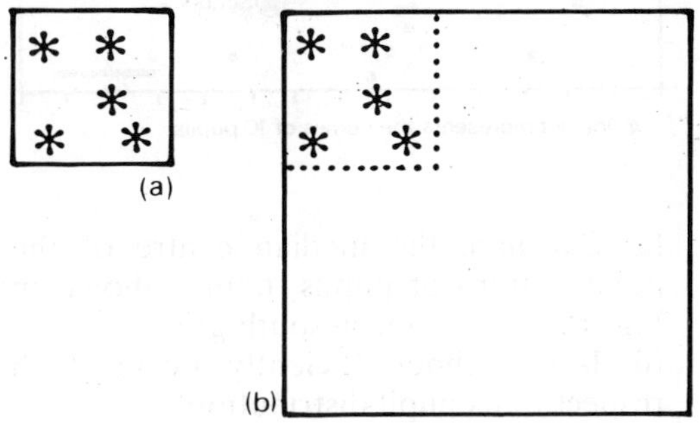

4 Exercises

Fig. 12 The hypothetical distribution of pupils attending a school

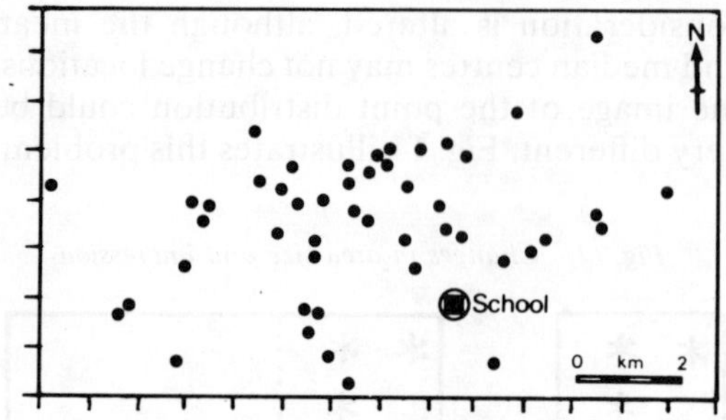

1 (a) Calculate the median centre of the point pattern of pupils' homes shown in Fig. 12. (Use a north–south grid.)
(b) Is the school efficiently located with respect to its pupil distribution?
(c) Impose a north-west/south-east grid upon the point pattern. Then calculate the median centre according to this different grid system. Is there any difference between this and the median centre calculated in (a) above? If so, what does this fact imply about the technique used here?

2 (a) A grid system is already marked on the map of the centre of Swanage (Fig. 13.) Mark on the x and y axes, and calculate the mean centre of different tourist perceptions of the location of the town centre.
(b) The crosses shown on Fig. 13 were obtained by survey at point A. Would you expect any difference in the results if the survey had been carried out at point B? If so, explain why, and suggest a likely location for the mean centre of this different distribution.

Table 5 Population density in wards of the London Borough of Brent, 1971

Density class (persons per hectare)	*Frequency (i.e., no. of wards) f*
20–49	8
50–79	12
80–109	4
110–139	2
140–169	3
170–199	1
200–229	1
	$\Sigma f = 31$

3 (a) Refer to Table 5 and calculate the approximate mean for the grouped data of population density in wards of the London Borough of Brent.
(b) Briefly describe the nature of this frequency distribution.
(c) Convert the data in the frequency column into cumulative percentages, and plot the cumulative frequency distribution for this data set.
(d) On the graph drawn in (c) above, mark in the estimated median value.

Fig. 13 Perceptions of the town centre of Swanage

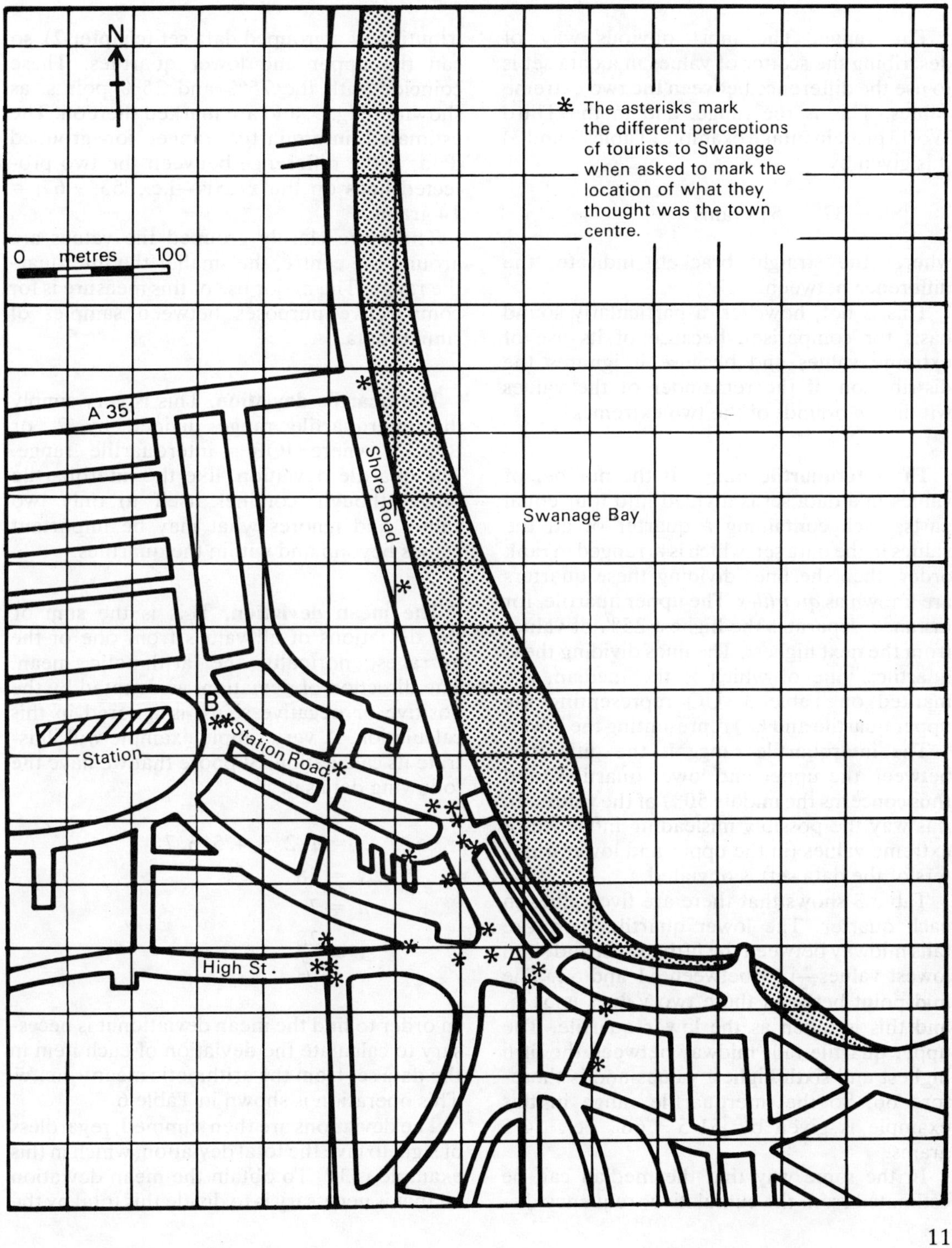

5 Measures of deviation, dispersion and variability

The range. The most obvious way of describing the scatter of values in a data set is to use the difference between the two extreme values. This is the range, and in the Third World protein intake example (Tables 2 and 3) it is given by:

$$|88 - 37| = 51 \text{ grams}$$

where: the straight brackets indicate 'the difference between . . . '.

This is not, however, a particularly sound basis for comparison because of its use of extreme values and because it ignores the distribution of the remainder of the values within the bounds of the two extremes.

The interquartile range. If the number of values in a data set is divided into four equal parts, each containing a quarter of all the values in the data set, which is arranged in rank order, then the lines dividing these quarters are known as *quartiles*. The upper quartile, for instance, separates the highest 25% of values from the next highest. The lines dividing these quartiles, one of which is the median, are marked on Table 3, UQ representing the upper quartile and LQ representing the lower.

The interquartile range is the difference between the upper and lower quartiles, and thus concerns the middle 50% of the values. In this way the possibly misleading influence of extreme values (in the upper and lower quarters of the data set) is obviated.

Table 3 shows that there are five values in each quarter. The lower quartile therefore falls midway between the fifth lowest and sixth lowest values—i.e., between 51 and 52. The mid-point between these two values is 51·5, and this is taken as the lower quartile. The upper quartile falls midway between the fifth highest and sixth highest values and is therefore 66. So the interquartile range in this example is given by $|51{\cdot}5 - 66|$, i.e., 14·5 grams.

In the same way that the median can be estimated from the cumulative frequency distribution of a grouped data set (chapter 2), so can the upper and lower quartiles. These coincide with the 75% and 25% points, as shown in Fig. 9, and are marked thereon. The estimated interquartile range, for grouped data, is the difference between the two projected lines on the x axis—i.e., $|53 - 67| =$ 14 grams.

The more closely grouped the values are around the centre, the smaller the interquartile range. The major use of this measure is for comparative purposes between samples of similar data.

The quartile deviation. This is very simply the interquartile range divided by 2, or IQR/2, where IQR = interquartile range. The quartile deviation, like the interquartile range, though, confines itself to only two values and ignores what may be important values beyond and within the quartiles.

The mean deviation. This is the sum of the deviations of all values from one of the averages, normally the arithmetic mean. The direction of deviation, as denoted by the positive or negative signs, is ignored in this calculation. A very simple example will illustrate its calculation. Suppose that we have the following data set:

$$x = 1, 2, 3, 4, 5, 6, 7$$

$$\therefore \Sigma x = 28$$

$$n = 7$$

$$\therefore \bar{x} = \frac{28}{7} = 4$$

In order to find the mean deviation it is necessary to calculate the deviation of each item in the data set from the arithmetic mean—i.e., $\bar{x}$. This operation is shown in Table 6.

The deviations are then summed, regardless of sign, to give the total deviation, which in this example is 12. To obtain the mean deviation all that is necessary is to divide this total by the

Table 6 Deviations from the mean in a hypothetical data set

Values in the data set x	Deviations from mean d or $\lvert x-\bar{x}\rvert$
1	3
2	2
3	1
4	0
5	1
6	2
7	3
	$\Sigma d = 12$

number of values in the data set, as given by

$$\frac{\Sigma d}{n} \text{ or } \frac{\Sigma|x - \bar{x}|}{n}$$

Here:

$$\frac{\Sigma|x - \bar{x}|}{n} = \frac{12}{7} = 1{\cdot}7$$

The same calculations for a more likely data set are shown in Table 7. This displays the same data as Table 3, but adds the deviations from the arithmetic mean, 60·2, in the second column. The list of deviations is summed to give a total deviation of 204·4. As there are 20 values in the sample (i.e., $n = 20$) the mean deviation is given by

$$\frac{204{\cdot}4}{20}$$

which equals 10·22 grams.

The mean deviation is not often used on its own, partly because it ignores the direction of deviation (positive or negative) and partly because it is normally desirable to emphasize the larger deviations by attaching more weight to them than to the smaller ones. One way of achieving the latter is to square the individual deviations as shown in the third column of Table 7 and as is done in the calculation of the standard deviation.

Table 7 Deviations and squared deviations from the mean protein intake—selected Third World countries, 1971

Protein intake x	Deviations from the arithmetic mean $\lvert x-\bar{x}\rvert$	Deviations squared $(x-\bar{x})^2$	Item squared x^2
88	27·8	772·84	7,744
81	20·8	432·64	6,561
80	19·8	392·04	6,400
69	8·8	77·44	4,761
66	5·8	33·64	4,356
66	5·8	33·64	4,356
66	5·8	33·64	4,356
64	3·8	14·44	4,096
64	3·8	14·44	4,096
58	2·2	4·84	3,364
54	6·2	38·44	2,916
54	6·2	38·44	2,916
53	7·2	51·84	2,809
52	8·2	67·24	2,704
52	8·2	67·24	2,704
51	9·2	84·64	2,601
50	10·2	104·04	2,500
50	10·2	104·04	2,500
49	11·2	125·44	2,401
37	23·2	538·24	1,369
	$\Sigma\lvert x-\bar{x}\rvert$ = 204·4	$\Sigma(x-\bar{x})^2$ = 3,029·2	Σx^2 = 75,510

The standard deviation. This gives the extreme values of any data set more weight by the use of a squaring procedure, which is also a mathematically acceptable way of overcoming the problem of direction of deviation. If we square the deviation from the arithmetic mean, the sign, positive or negative, is converted to a positive value, since a negative value multiplied by a negative value becomes positive.

The total squared deviation is given by the formula $\Sigma(x - \bar{x})^2$. The mean squared deviation is therefore $\frac{\Sigma(x - \bar{x})^2}{n}$, which is commonly known as the variance of the data set, or σ^2 (Greek letter sigma). This is the average of the squares of the deviations from the arithmetic mean.

The calculations to find the variance of the simple hypothetical data set used earlier are shown in Table 8.

Table 8 Calculations for the variance of a hypothetical data set

x	$(x - \bar{x})$	$(x - \bar{x})^2$
1	−3	9
2	−2	4
3	−1	1
4	0	0
5	+1	1
6	+2	4
7	+3	9
		$\Sigma(x - \bar{x})^2 = 28$

If $\Sigma(x - \bar{x})^2 = 28$,
then

$$\frac{\Sigma(x - \bar{x})^2}{n} = \frac{28}{7} = 4$$

$$\therefore \quad \sigma^2 = 4$$

The variance, however, gives a value that is positive, whereas a deviation value should summarize differences from the mean in both positive and negative directions. This can be done by finding the square root of the variance, as the square root of any number has both a positive and a negative value. This square root of the variance is known as the standard deviation, and is shown as σ, since the square root of σ^2 is σ.

So,

$$\sigma = \sqrt{\frac{\Sigma(x - \bar{x})^2}{n}}$$

In the case cited above:

$$\sigma = \sqrt{4} = \pm 2$$

So, the standard deviation is the square root of the average of the squares of the deviations from the arithmetic mean of the data set!

Before calculating the standard deviation of a sample data set, it would be worthwhile summarizing the steps involved in the calculation of the standard deviation:

1 Total x	1 Σx
2 Find arithmetic mean of all x	2 $\frac{\Sigma x}{n} = \bar{x}$
3 Find deviations from the mean of each value	3 $(x - \bar{x})$
4 Square deviation from the mean of each value	4 $(x - \bar{x})^2$
5 Total the squared deviations	5 $\Sigma(x - \bar{x})^2$
6 Find the arithmetic mean of the squared deviations by dividing by n (i.e., find the variance)	6 $\sigma^2 = \frac{\Sigma(x - \bar{x})^2}{n}$
7 Calculate the square root of the variance	7 $\sigma = \sqrt{\frac{\Sigma(x - \bar{x})^2}{n}}$

The third column of Table 7 shows the squared deviations from the mean protein intake of selected Third World countries in 1971. The total of these is 3,029·2, which leads to the fifth step in the calculation of the standard deviation. Thus:

$$\Sigma(x - \bar{x})^2 = 3{,}029{\cdot}2$$

$$\therefore \quad \frac{\Sigma(x - \bar{x})^2}{n} = \frac{3{,}029{\cdot}2}{20} = 151{\cdot}46 = \sigma^2$$

$$= \text{the variance}$$

$$\therefore \quad \sqrt{\frac{\Sigma(x - \bar{x})^2}{n}} = \sqrt{151{\cdot}46} = \pm 12{\cdot}31 \text{ grams}$$

$$= \sigma = \text{the standard deviation}$$

Clearly, this method of calculation of the standard deviation is lengthy and time-consuming, particularly if a calculator is not available. Mathematically, though, the standard deviation is the best method of describing the scatter of values in a data set. It is also required as a basis of many other statistical techniques. It is, therefore, a very necessary and important statistical measure, and a shorter method of calculation would be very useful. Fortunately, it is possible to manipulate algebraically the formula for the standard deviation given above to a form that requires fewer tedious additions and subtractions:

$$\sigma = \sqrt{\frac{\Sigma x^2}{n} - \bar{x}^2}$$

Details of this manipulation are given in *Statistical Methods and the Geographer*, by S. Gregory (Longman, 1963).

The fourth column in Table 7 gives the squares of the grams of protein intake for each country in the sample, and the sum of these equals 75,510. By the alternative formula, then:

$$\sigma = \sqrt{\frac{75{,}510}{20} - 60{\cdot}2^2}$$

$$\sigma = \sqrt{3775{\cdot}5 - 3624{\cdot}04}$$

$$\sigma = \sqrt{151{\cdot}46}$$

$$\therefore \sigma = \pm 12{\cdot}31 \text{ grams}$$

Grouped data. Once again, an estimate for the mean and standard deviations can be obtained from grouped data if the original data is unavailable. The principle and procedure behind this estimate are the same as those behind the calculation of the estimated grouped mean (chapter 3). Table 9 lists the information required in the calculation of the estimated grouped mean deviation, and the formula for the estimation is:

$$\frac{\Sigma f|x_0 - \bar{x}|}{n}$$

where: $\bar{x}$ = the estimated grouped mean (which in this example is 60·5 grams);
f = the number of items in each class, or the frequency;
$|x_0 - \bar{x}|$ = the deviation of each mid-point value from the estimated grouped mean;
n = the total number of values in the data set.

The deviation of each mid-point value from the estimated grouped mean, 60·5, (see chapter 3) is calculated, and these values are then multiplied by the frequency (or number of items in each class or group). These products are summed and divided by n. For the example given here the calculations are detailed below:

$$\text{estimated mean deviation} = \frac{\Sigma f|x_0 - \bar{x}|}{n}$$

$$= \frac{192{\cdot}0}{20}$$

$$= 9{\cdot}6 \text{ grams}$$

This is to be compared with the actual mean deviation of 10·27 grams.

Table 9 Calculation of the estimated mean and standard deviation from the grouped protein intake data

For both calculations				*For mean deviation*	*For the standard deviation*	
A *Class (gm of protein)*	B *Mid-point value* x_0	C *Deviation of* x_0 *from* $\bar{x}$ *(i.e., 60·5)*	D *Frequency (f)*	E $\therefore$ $f\|x_0-\bar{x}\|$	F *Deviation squared* $(x_0-\bar{x})^2$	G $\therefore$ $f(x_0-\bar{x})^2$
30–39	34·5	26·0	1	26·0	676	676
40–49	44·5	16·0	1	16·0	256	256
50–59	54·5	6·0	9	54·0	36	324
60–69	64·5	4·0	6	24·0	16	96
70–79	74·5	14·0	0	0	196	0
80–89	84·5	24·0	3	72·0	576	1,728
			$\Sigma f = 20$	$\Sigma = 192{\cdot}0$		$\Sigma = 3{,}078$

The formula for the estimated grouped standard deviation is:

$$\sigma = \sqrt{\frac{\Sigma f(x_0 - \bar{x})^2}{n}}$$

Column G in Table 9 contains the information necessary for the calculation, and in this case:

$$\sigma = \sqrt{\frac{3{,}078}{20}}$$

$$\sigma \doteq \sqrt{153{\cdot}9}$$

$$\therefore \sigma = \pm 12{\cdot}41 \text{ grams}$$

This is to be compared with an actual standard deviation of ±12·31 grams. Although the difference here is not great, it should be repeated that the standard deviation for the grouped data is only an estimate. It does, however, have its uses, particularly in curtailing the time-consuming nature of the calculations in the standard deviation.

The standard deviation and the normal distribution. The frequency curve of the distribution of any data set can be described by its arithmetic mean and by its standard deviation. If we can assume that the frequency distribution is normal (see chapter 1) then it is possible to postulate the number of occurrences or items at any given value or between given values, if the arithmetic mean and standard deviation are known.

Assuming, then, a normal distribution, that is one whose frequency curve is centred on the arithmetic mean and is symmetrical about that mean, we can use the laws of probability to state that:

between $+1\sigma$ and -1σ of $\bar{x}$, will be found 68·27% of all the items in the data set;
between $+2\sigma$ and -2σ of $\bar{x}$, will be found 95·45% of all the items in the data set;
between $+3\sigma$ and -3σ of $\bar{x}$, will be found 99·7% of all the items in the data set;
between $+4\sigma$ and -4σ of $\bar{x}$, will be found 99·99% of all the items in the data set.

Put another way, this means that 68% (figure rounded off) of all the values in the data set should lie within 1σ of the arithmetic mean of that data set. Fig. 14 illustrates this idea of probability and the normal distribution. This relationship between probability and the normal distribution is fundamental, for the size of many samples to be taken is determined by this basic concept. To be exact the distribution should correctly be called the 'normal probability distribution'.

Fig. 14 Probability and the normal distribution

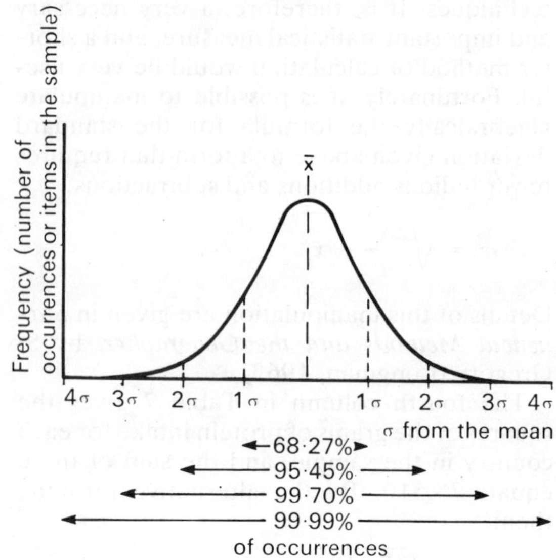

From this relationship we can say that a difference of more than or less than 3 standard deviations from the arithmetic mean is only to be expected once in every 300 items or cases (99·7% occur within 3σ, thus 0·3% are outside 3σ, and 0·3% = 0·3 in 100 ≏ 1 in 300) and that all items in the sample are normally contained within 4 standard deviations of the mean. This is not always so, as one in every 10,000 cases should be found outside this limit (i.e., 0·01%). We can use this knowledge of probability theory and the normal distribution when examining data from small samples. So if in a sample data set of 50 items one of the values exceeds ±3 standard deviations from the mean of that data set, then we could regard it with just a little suspicion. It may of course be perfectly acceptable, as the one case in every 300 will occur at some point, but it should be checked as it could mean that a thermometer has developed a fault, a figure has been misread, or some other instrument broke down for a short time.

The newcomer to probability theory should test the data set of protein intake of selected Third World countries against the normal distribution. Table 10 sets out the number and

Table 10 The frequency distribution of the protein intake data, as defined by the standard deviation and the mean

Range, as defined by standard deviation	Actual values (grams)	Number of occurrences	Percentage of occurrences
More than $+3\sigma$	More than 97·13	0	0
From $+2\sigma$ to $+3\sigma$	84·82–97·13	1	5
From $+1\sigma$ to $+2\sigma$	72·51–84·82	2	10
From $\bar{x}$ to $+1\sigma$	60·20–72·51	6	30
From -1σ to $\bar{x}$	47·89–60·20	10	50
From -2σ to -1σ	35·58–47·89	1	5
From -3σ to -2σ	23·27–35·58	0	0
Less than -3σ	Less than 23·27	0	0

percentages of values occurring within the ranges defined by the standard deviation and the mean. It is interesting to compare the frequency distribution of this sample data set with that of a normally distributed data set. In this case there seems to be some conformity to the normal distribution. If it is known, however, that the distribution is skewed, then the three standard deviations check cannot be relied upon to find possible faults. Information regarding the possible skewness of this data set is not available at first hand, although, intuitively—simply because it only concerns the less developed nations (i.e., a biased and deliberately selected sub-total of the whole)—it seems likely that there may be some skewness. Therefore great reliability cannot be placed upon such a small sample data set.

Samples of pebble size in a particular location, or of altitudes at random points, or similar data concerning the natural world should fit the normal distribution rather more closely.

The following three indices of scatter can also be useful, particularly for the purposes of direct comparison between two seemingly similar data sets.

The index of variability. This expresses the quartile deviation as a percentage of the median.

$$V_i = \frac{\text{Quartile deviation}}{\text{median}} \cdot 100\%^*$$

In the example used here:

$$V_i = \frac{7{\cdot}25}{56} \cdot 100 = 12{\cdot}95\%$$

Relative variability. This expresses the mean deviation as a percentage of the arithmetic mean.

$$V_r = \frac{\text{Mean deviation}}{\text{Arithmetic mean}} \cdot 100\%^*$$

$$= \frac{\frac{\Sigma|x - \bar{x}|}{n}}{\bar{x}} \cdot 100\%$$

or

$$= \frac{\Sigma|x - \bar{x}|}{n} \cdot \frac{1}{\bar{x}} \cdot 100$$

In this example:

$$V_r = \frac{10{\cdot}22}{60{\cdot}2} \cdot 100 = 16{\cdot}98\%$$

Coefficient of variation. This expresses the standard deviation as a percentage of the arithmetic mean.

$$V_c = \frac{\text{Standard deviation}}{\text{Arithmetic mean}} \cdot 100\%$$

$$= \frac{\sigma}{\bar{x}} \cdot 100\%$$

* The point in the equation represents the notation used in this book for multiplication.

In this example:

$$V_c = \frac{12 \cdot 31}{60 \cdot 2} \cdot 100 = 20 \cdot 45\%$$

With all three indices above, the higher the percentage, the greater the variability. The major use of these three indices lies in comparison between data sets, particularly where the magnitude of the values differs between two or more samples. For example, two samples of pebble size, taken at different points on a beach profile, gave the following results: at point A, higher up the beach, the mean pebble length was found to be 3·6 cm, with a standard deviation of 1·02 cm; at point B, lower on the profile, the mean length was 1·02 cm with a standard deviation of 0·29 cm. From these results it appears that there is much greater variation at point A, but if the standard deviation of each sample is related to its arithmetic mean, as in the coefficient of variation, the variations are shown to be very similar:

for point A

$$V_c = \frac{1 \cdot 02}{3 \cdot 6} \cdot 100 = 28 \cdot 33\%$$

for point B

$$V_c = \frac{0 \cdot 29}{1 \cdot 02} \cdot 100 = 28 \cdot 43\%$$

In other words, the variations in the observations for both samples are on different scales, but when related to the magnitude of the mean of each data set, the variations are very similar.

6 Summary of mathematical notation and formulae

1 Mathematical symbols

x the values (items or variables)
x_0 the mid-point value of each class for grouped data sets
n the total number of values in the data set
Σ the sum of
$\bar{x}$ the arithmetic mean (μ)
f the frequency of

2 Formulae

(a) The arithmetic mean $= \bar{x} = \dfrac{\Sigma x}{n}$

(b) The grouped mean $= \dfrac{\Sigma f x_0}{n}$

(c) The quartile deviation

$$= \frac{\text{interquartile range}}{2}$$

(d) The quartile coefficient of dispersion or deviation

$$= \frac{\text{the quartile deviation}}{\text{the average of the two quartiles}}$$

(e) The mean deviation $= \dfrac{\Sigma|x - \bar{x}|}{n}$

(f) The mean deviation for grouped data

$$= \frac{\Sigma f|x_0 - \bar{x}|}{n}$$

(g) The variance $= \sigma^2 = \dfrac{\Sigma(x - \bar{x})^2}{n}$

(h) The standard deviation

$$= \sigma = \sqrt{\frac{\Sigma(x - \bar{x})^2}{n}} = \sqrt{\frac{\Sigma x^2}{n} - \bar{x}^2}$$

(i) The standard deviation for grouped data

$$= \sqrt{\frac{\Sigma f(x_0 - \bar{x})^2}{n}}$$

(j) The index of variability

$$= V_i = \frac{\text{quartile deviation}}{\text{median}} \cdot 100\%$$

(k) Relative variability

$$= V_r = \frac{\text{mean deviation}}{\text{arithmetic mean}} \cdot 100\%$$

$$= \frac{\frac{\Sigma|x - \bar{x}|}{n}}{\bar{x}} \cdot 100\%$$

$$= \frac{\Sigma|x - \bar{x}|}{n} \cdot \frac{1}{\bar{x}} \cdot 100\%$$

(l) Coefficient of variation

$$= V_c = \frac{\text{standard deviation}}{\text{arithmetic mean}} \cdot 100\%$$

$$= \frac{\sigma}{\bar{x}} \cdot 100\%$$

7 Exercises

1 Put the data given in Table 11 into rank order.

2 Find the following measures of deviation for the data in Table 11:
(a) the range;
(b) the interquartile range;
(c) the quartile deviation;
(d) the mean deviation;
(e) the standard deviation;
(f) the index of variability;
(g) the relative variability;
(h) the coefficient of variation.

3 (a) How closely does the distribution of values around the arithmetic mean conform to the normal probability distribution?
(b) Suggest ways in which bias, leading to a skewed distribution, may have entered the sample.

4 Compare the measures of central tendency and of statistical deviation of the data given in Table 11 with the same measures of the data given in Table 2.

Table 11 Protein intake for selected countries of the developed world, 1971

Country	*Protein intake per person per day,* 1971
	grams
Australia	92
Belgium	88
Canada	95
Denmark	89
Eire	94
France	101
Greece	99
Israel	89
Italy	85
Japan	75
Netherlands	85
New Zealand	107
Norway	81
Portugal	83
Spain	82
Sweden	80
Switzerland	88
UK	88
USA	96
West Germany	81

$\Sigma x = 1778$

$\bar{x} = 88{\cdot}9$

median = 88

mode = 88

$n = 20$

8 The significance of the difference between samples

The two data sets of protein intake (Tables 2 and 11) are very clearly different from each other, but there are many cases of data sets where differences do exist but are not so clear at first sight.

In order to test the significance of difference between two sample data sets we ask whether the differences between them could have occurred purely by chance or whether the differences in the samples really do reflect differences in the populations from which they have been drawn. This necessarily involves the concept of sampling error, the determination of which is based upon probability theory.

Probability theory. If the full set of data is considered (i.e., not just a sample) then there is 100% probability that any individual item or occurrence will lie between or be equal to the values of the outer limits of that full data set. In other words, the full data set must contain all the items in that data set and therefore all the probabilities.

If it is known that the frequency distribution is normal, then we can also state that:

the probability (p) of an item being $\geqslant \bar{X}$
= 50% or 0·5

(where $\bar{X}$ = the arithmetic mean of the full data set),

and that:

the probability (p) of an item being $\leqslant \bar{X}$
= 50% or 0·5

therefore the total probability
= 100% or 1·0

Also, assuming a normal distribution, we can state that:

pX (i.e., the probability of any item) differing from $\bar{X}$ by $\leqslant 2\sigma$
= 95·45% or 0·9545

This means that there is a 95·45% probability that any item, X, will occur within two standard deviations of $\bar{X}$.

pX differing from $\bar{X}$ by $\geqslant 2\sigma$
= 4·55% or 0·0455

pX differing from $\bar{X}$ by $\leqslant 3\sigma$
= 99·7% or 0·997

pX differing from $\bar{X}$ by $\geqslant 3\sigma$
= 0·3% or 0·003

Levels of statistical significance. So, it can be simply said that any item, X, in any data set can be given a probability for taking on any value. This probability within a data set can be viewed as a scale of likelihood, ranging from 0% to 100%. It is probability levels along this scale of likelihood which form the basis of most scientific proof—i.e., very few facts can be proved as being 100% true, but most ideas have a certain likelihood or probability that they are true. The levels of probability accepted by the scientific community are those defined by the arithmetic mean and the standard deviation, as above, such that the existence of a difference, a relationship or a fact can be proved to the following levels of statistical significance:

95·45%; 99·7%; 99·99%

Partly for the sake of convenience, the levels of statistical significance used for many statistical tests may often be expressed in different ways. For instance:

$p = 0{\cdot}05$ or the 95% level
$p = 0{\cdot}01$ or the 99% level
$p = 0{\cdot}001$ or the 99·9% level

When we state that $p = 0{\cdot}05$, this means that we have obtained a result that would only arise once every 20 times (5 in every 100) on the basis of pure chance.

These levels of statistical significance are dealt with at greater length later in chapter 10, Hypothesis testing; and chapter 12, Correlation.

Sampling error. Of necessity, we are concerned largely with sample data sets rather than complete data sets, and we have no guarantee that the parameters of the sample ($\bar{x}$ and σ), are in fact the true parameters of the complete population of items. It is the nature of any sample that it will contain some error, and so too will its parameters.

This error can be estimated by applying a formula that gives the standard error of the mean. The word 'standard' comes simply from the fact that the standard deviation is used in its calculation.

Standard error of the mean. This is found by dividing the sample standard deviation by the square root of the number of items in the sample:

$$\text{S.E. } \bar{x} = \frac{\sigma}{\sqrt{n}}$$

where:

$\bar{x}$ = the arithmetic mean of the sample data set
n = the number of items in the sample
σ = the sample standard deviation

Now, if the data is normally distributed, there is a 68% probability that any value will occur within a range of $\pm 1\sigma$ of $\bar{X}$, a 95% probability that any value will occur within $\pm 2\sigma$ of $\bar{X}$, and a 99·7% probability that any value will occur within $\pm 3\sigma$ of $\bar{X}$.

It follows, therefore, that the true arithmetic mean of any complete data set, $\bar{X}$, will be equal to the sample mean, $\bar{x}$, ±1, 2 or 3 standard errors of the mean with respectively 68%, 95% and 99·7% levels of probability. Or:

$$\bar{X} = \bar{x} \pm \frac{\sigma}{\sqrt{n}}$$ with a 68·27% probability

$$\bar{X} = \bar{x} \pm \frac{2\sigma}{\sqrt{n}}$$ with a 95·45% probability

$$\bar{X} = \bar{x} \pm \frac{3\sigma}{\sqrt{n}}$$ with a 99·7% probability

An actual example should help to illustrate the meaning of the standard error of the mean. Table 12 shows the percentage of arable land in two random samples, each of 12 farms, which were selected in two areas of contrasted soil types.

Table 12 The percentage of arable land on farms in two areas

Area 1	**Area 2**
16%	32%
21	31
18	35
25	29
14	33
28	29
19	34
26	31
22	36
16	32
21	29
19	33
$\Sigma_x = 245$	$\Sigma_y = 384$
$n_x = 12$	$n_y = 12$
$\bar{x} = 20{\cdot}42\%$	$\bar{y} = 32{\cdot}0\%$
$\sigma_x = \pm 4{\cdot}11\%$	$\sigma_y = \pm 2{\cdot}24\%$

For the two areas shown the standard errors of the means are as follows:

Area 1:

$$\text{S.E. } \bar{x} = \frac{4{\cdot}11}{\sqrt{12}} = \frac{4{\cdot}11}{\pm 3{\cdot}46} = \pm 1{\cdot}19$$

Area 2:

$$\text{S.E. } \bar{y} = \frac{2{\cdot}24}{\sqrt{12}} = \frac{2{\cdot}24}{\pm 3{\cdot}46} = \pm 0{\cdot}65$$

This means that the true arithmetic mean is equal to the sample arithmetic mean plus or minus 1·19 (for Area 1) to a 68% level of probability. Or, for Area 1:

$\bar{X}$ ranges from 19·23% to 21·61% to a 68% level of probability;

and that:

$\bar{X}$ ranges from 18·04% to 22·80% to a 95% level of probability;

$\bar{X}$ ranges from 16·85% to 23·99% to a 99·7% level of probability.

For Area 2 the corresponding values are:

31·35% to 32·65%
30·70% to 33·30%
30·05% to 33·95%

Standard error of the difference. This is used to show the significance of the difference between two sample data sets. It is a function of the standard errors of the means of both sample data sets and of the difference between these means. (The difference between the sample means is also liable to some error as well as the means themselves.) The question we ask ourselves is whether or not the differences between the sample means are sufficiently great to justify a conclusion that the true means also differ significantly.

The calculation of the standard error of the difference depends on the fact that the standard error of the mean is a function of the standard deviation. The standard deviation is the square root of the variance, and it is the variance between the two sample data sets that we are testing.

As the standard error of the mean is

$$\frac{\sigma}{\sqrt{n}},$$

then the variance of the sample mean is the standard error squared—i.e.,

$$\left(\frac{\sigma}{\sqrt{n}}\right)^2 \quad \text{or} \quad \frac{\sigma^2}{n}.$$

If, then, we are comparing a sample of n_x items drawn from a population whose standard deviation is σ_x with a sample of n_y items drawn from a population whose standard deviation is σ_y, the variance of the difference of sample means of $\bar{x}$ and $\bar{y}$ is given by:

$$\text{Var.}\,|\bar{x} - \bar{y}| = \frac{{\sigma_x}^2}{n_x} + \frac{{\sigma_y}^2}{n_y}$$

or:

> the variance of the difference between $\bar{x}$ and $\bar{y}$ = the variance of $\bar{x}$ *plus* the variance of $\bar{y}$

But this is the variance of the difference, and so the formula for the standard error of the difference is the square root of this variance—i.e.:

$$\text{S.E.}\,|\bar{x} - \bar{y}| = \sqrt{\frac{{\sigma_x}^2}{n_x} + \frac{{\sigma_y}^2}{n_y}}$$

Providing that the data set is normally distributed, the standard error of the difference between $\bar{x}$ and $\bar{y}$ has the same probability characteristics as the normal frequency curve, as does the standard error of the mean. So the probability of the actual difference between the sample means being more than twice this standard error of the difference is nearly 5%; and the probability of it being more than three times the standard error of the difference is only 0·3%.

In other words, if the actual difference between $\bar{x}$ and $\bar{y}$ (the sample means) is greater than twice the standard error of the difference, then it is unlikely, though not completely certain, that a difference of this size between the two sample means occurred by chance. So the difference can be labelled as 'probably significant'. If the actual difference between $\bar{x}$ and $\bar{y}$ is greater than three times the standard error of the difference, then the difference between these means can be labelled as 'highly significant' (99·7% certain).

Again, an example will help to illustrate both the calculation and the use of the standard error of the difference. The standard error of the difference between the percentage of arable land in farms in Areas 1 and 2, shown in Table 12, is calculated below:

$$\begin{aligned}\text{S.E.}\,|\bar{x} - \bar{y}| &= \sqrt{\frac{4{\cdot}11^2}{12} + \frac{2{\cdot}24^2}{12}} \\ &= \sqrt{\frac{16{\cdot}89}{12} + \frac{5{\cdot}02}{12}} \\ &= \sqrt{1{\cdot}41 + 0{\cdot}42} \\ &= \sqrt{1{\cdot}83} \\ &= \pm 1{\cdot}35\end{aligned}$$

Now, this standard error of the difference is compared with the actual difference between the means of both samples. The actual difference is given by:

$$\begin{aligned}|\bar{x} - \bar{y}| &= |20{\cdot}42 - 32{\cdot}0| \\ &= 11.58\end{aligned}$$

It can be seen that the actual difference of 11·58 is more than three times greater than the standard error of the difference of 1·35. The difference between the two samples can therefore be described as highly significant, and it is highly unlikely that the difference between them occurred by chance.

If the actual difference between the means had been less than two times the standard error of the difference, then this would have meant that there is less than a 95% probability that the difference between the two samples is significant. This does not mean, however, that there is definitely no difference between the two—simply that the difference has not been statistically proved. In fact, the very concept of levels of statistical significance should be treated with much caution. The real world is somewhat more complicated than these levels would suggest, and they should not be applied if the number of items in the two samples totalled together is less than 30. This is simply on the grounds that the greater the number of items in the sample, the greater the reliability of that sample.

Of course, this particular method of assessing the significance of difference (i.e., the standard error of the difference being twice or thrice the actual difference) is a little rough and ready. So, it is possible to use a refinement of the standard error of the difference. This is called the Student's *t* test, and gives a more accurate measure of the difference. It is also more reliable with samples of less than 30 items.

Student's *t* test. Instead of looking to see if the difference is greater than two or three standard errors, the Student's *t* is a calculation which expresses exactly how many times greater it is. This measure is then referred to prepared tables or to a graph, from which the level of statistical significance of the difference can be assessed.

$$t = \frac{\text{the actual difference between the sample means}}{\text{the standard error of the difference}}$$

$$= \frac{|\bar{x} - \bar{y}|}{\sqrt{\frac{\sigma_x^2}{n_x} + \frac{\sigma_y^2}{n_y}}}$$

The meaning of the value so obtained will depend upon the number of items in the samples under study. It is therefore necessary to calculate the *degrees of freedom* allowed by this number. These are given by:

$$df = (n_x - 1) + (n_y - 1)$$

This gives a value representing the number of items on which the comparison is based. The smaller the number of items in the samples, the less the degrees of freedom, and therefore the larger the value *t* must be in order to be classed as showing a significant difference.

Using the data sets shown in Table 12, and for which the standard error of the difference has already been calculated, the Student's *t* test gives the following results:

$$t = \frac{11{\cdot}58}{1{\cdot}35} = 8{\cdot}57 \qquad df = (12 - 1) + (12 - 1) = 22$$

These values are then referred to the *Cambridge Elementary Statistical Tables* (Lindley and Miller, 1970, Table 3) or to the graph prepared from these tables. Such a graph is shown in Fig. 15. As can be seen from this graph, our value of *t* occurs above the 0·1% line, this signifying a difference that is

Fig. 15 Student's t graph

(Student's *t* Graph based on data in Cambridge Elementary Statistical Tables and reproduced here from Gregory, S., *Statistical Methods and the Geographer*, Longman, 1963.)

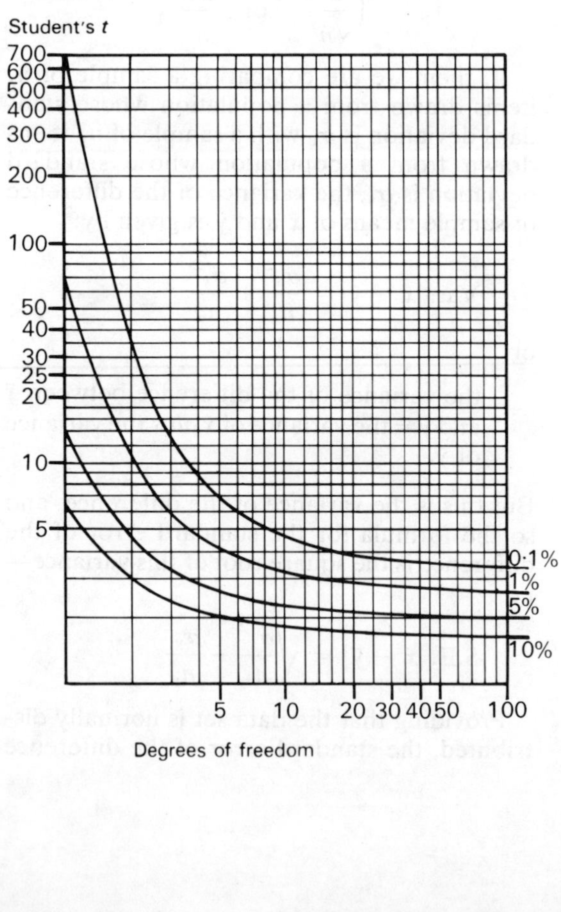

'highly significant'. Had this value occurred between the 0·1% and 1% lines, it would have indicated a difference that could be described as 'significant', whereas a value between the 1% and 5% lines would describe a difference that was 'probably significant'. These percentage lines indicate the percentage probability that the difference between the samples could have occurred by chance.

Dispersion diagrams. Although not a statistically sound indication of the difference between samples, these are a simple visual guide to the usefulness of further analysis. The scatter of values in each sample is plotted, using the same scale for each sample. For each sample data set the median and the lower and upper quartiles are marked on the scatter.

Conclusions drawn from these diagrams depend upon the relative positions of the upper and lower quartiles and medians. Three hypothetical examples illustrate these conclusions in Fig. 16. The first situation, in Fig. 16(a), shows that the interquartile ranges of data sets P and Q do not overlap—or the lower quartile of Q is above the upper quartile of P. This suggests a difference that is approximately equivalent to that shown at the highest level of probability, and which can therefore be described as 'highly significant'. It must be remembered, however, that these dispersion diagrams only give a rough 'rule of thumb' method of analysis. The term 'highly significant' should therefore be applied only if the context of dispersion diagrams is made clear. They are in fact normally only used as a preliminary guide to the usefulness of further statistical analysis.

Fig. 16(b) shows that the lower quartile of data set S is above the median but below the upper quartile of data set R. In this case the difference is 'probably significant'.

Fig. 16(c) illustrates the case where no significant difference between data sets X and Y is demonstrated. Here, the lower quartile of Y is above the lower quartile but below the median of X.

An actual example will help to illustrate to the reader the use of this technique. Fig. 17 shows the infant mortality rates (per '000, 1966) of three groups of selected countries: A, Africa and the Middle East; B, South and Central America; C, Europe and North

Fig. 16 Dispersion diagrams of 3 hypothetical situations

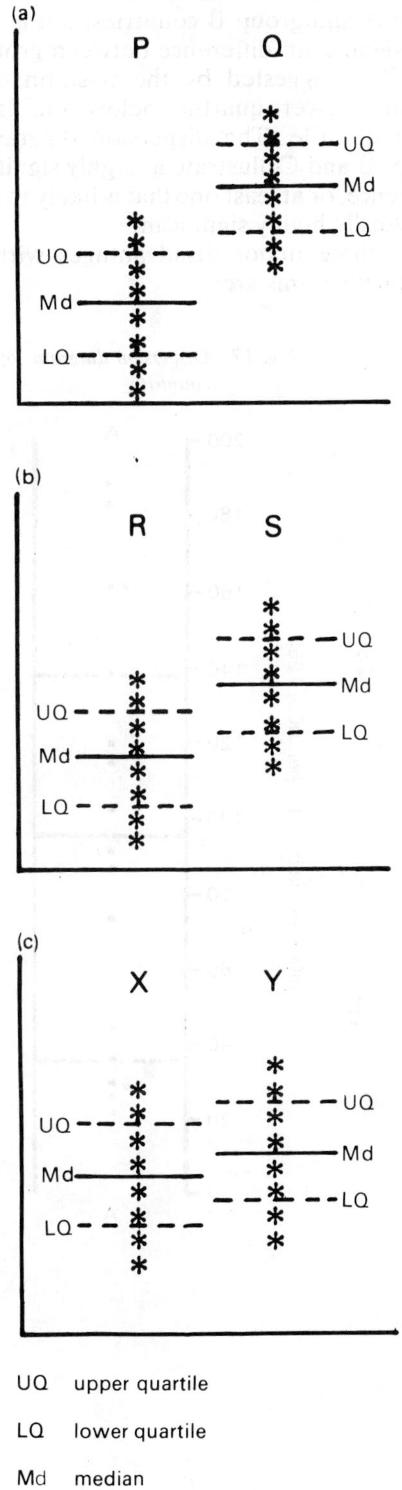

America. There is no significant difference shown between the infant mortality rates of group A and group B countries, but a 'probably significant' difference between groups A and C is suggested by the position of the former's lower quartile below the latter's upper quartile. The dispersion diagrams of groups B and C illustrate a 'highly significant' difference, or at least one that is likely to prove statistically highly significant.

The three major disadvantages with dispersion diagrams are:

(a) they can only be approximate as no statistical probability can be attached to the terms 'probably significant' and 'highly significant';

(b) they do not allow for differences in sample size which, statistically, it is unwise to ignore;

(c) the median and upper and lower quartiles can often be inaccurate or misleading.

Despite these drawbacks, they are simple and quick to construct and yield a clear visual result.

Fig. 17 Dispersion diagrams of the infant mortality rates of three groups of selected countries

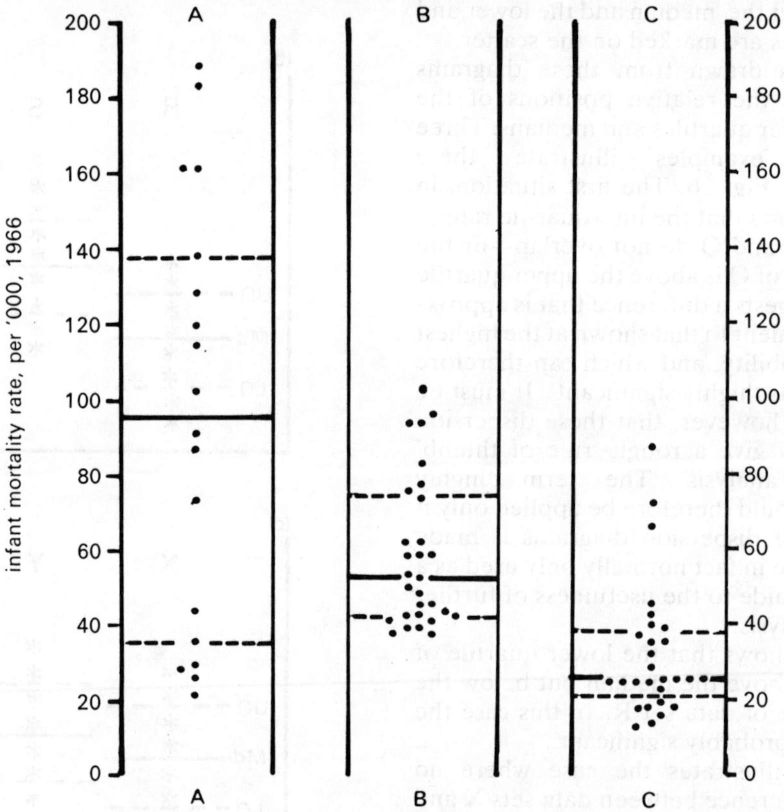

9 Exercises

Table 13 shows the kilometres of rail route per 10,000 population (for 1966) for countries in three different regions of the world. Relevant statistical measures for each data set are also given.

Table 13 Kilometres of rail route per 10,000 population, 1966

A Africa and the Middle East		B South and Central America		C Europe and North America	
Country	*Index*	*Country*	*Index*	*Country*	*Index*
Ethiopia	0·43	Venezuela	0·83	Romania	0·50
Nigeria	0·49	Panama	0·84	Netherlands	2·57
Cameroun	0·91	Trinidad and Tobago	1·70	Greece	2·96
Kenya	0·93	Colombia	1·79	Portugal	3·81
Senegal	1·06	Peru	2·14	Italy	4·02
Malawi	1·11	Guyana	2·22	UK	4·36
Ghana	1·16	Jamaica	2·24	Belgium	4·68
Morocco	1·24	Nicaragua	2·26	Spain	5·89
Iran	1·35	El Salvador	2·40	West Germany	5·93
Mali	1·36	Ecuador	2·43	Yugoslavia	5·94
Malagasy Republic	1·39	Guatemala	2·46	Bulgaria	7·01
UAR	1·41	Dominican Rep.	3·71	France	7·65
Lebanon	1·46	Brazil	4·13	Hungary	8·09
Uganda	1·46	Honduras	4·71	Irish Republic	8·13
Ivory Coast	1·63	Costa Rica	5·05	Poland	8·42
Syria	1·72	Mexico	5·18	Denmark	8·43
Guinea	1·79	Paraguay	5·31	Switzerland	8·45
Jordan	1·87	Cuba	7·44	Austria	9·01
Sierra Leone	2·05	Chile	9·41	Czechoslovakia	9·23
Dahomey	2·31	Bolivia	9·42	East Germany	9·47
Iraq	2·44	Uruguay	11·15	Luxembourg	10·09
Zambia	2·55	Argentina	18·63	Finland	11·50
Israel	2·59			Norway	11·52
Liberia	3·05			USA	17·32
Tanzania	3·09			Sweden	17·67
Zaire (CDR in 1966)	3·16			Canada	46·82
Algeria	3·30				
Sudan	3·76				
Tunisia	4·52				
Mozambique	5·08				
Swaziland	5·87				
Congo (Rep.)	6·00				
Angola	6·26				
Mauritania	6·43				
Rhodesia	7·40				
Botswana	10·79				
South Africa	11·32				
$n=37$	$\Sigma x = 114{\cdot}74$, $\bar{x} = 3{\cdot}10$, $\sigma = \pm 2{\cdot}62$	$n=22$	$\Sigma x = 105{\cdot}45$, $\bar{x} = 4{\cdot}79$, $\sigma = \pm 4{\cdot}14$	$n=26$	$\Sigma x = 239{\cdot}47$, $\bar{x} = 9{\cdot}21$, $\sigma = \pm 8{\cdot}47$

Compiled from the Oxford Economic Atlas of the World, 1974

1 (a) Plot the information shown for data sets A, B and C on an arithmetic graph paper (Table 13).
(b) For each data set mark in the median and upper and lower quartiles.
(c) What do the results of (a) and (b) indicate about the differences between these data sets?

2 (a) Calculate the standard errors of the difference for the following pairs of data sets: A and B; B and C; A and C.
(b) From the results in (a) what can we state about the differences between these regions and their rail density per head of population?
(c) What conclusions can be made regarding the differing levels of development of the three regions?

3 (a) Calculate the Student's t statistic for the pairs of data sets already used.
(b) Calculate the degrees of freedom allowed for each pair and then refer your statistic to the Student's t graph.
(c) Do the results support the findings in 1 and 2 above?

10 Hypothesis testing

Parametric and non-parametric statistics. All of the statistical tests and techniques so far discussed have been parametric. This means that from the sample data set the test makes assumptions about the nature of the full population from which the values were taken—usually the assumption of the normal distribution. This covers all the statistical tests that use the parameters of the arithmetic mean and standard deviation to describe the nature of the frequency distribution. Parametric tests also normally involve data in absolute numbers or values rather than ranks.

Non-parametric tests do not make a series of assumptions about the data under analysis, and are often more applicable to geographic problems. They are sometimes referred to as distribution-free statistical tests, and the data are generally ranked or grouped.

Before considering a non-parametric test in more detail, it is necessary to explain how to set up a hypothesis that can be subjected to such a test.

Hypothesis testing and statistical inference. Geographers, and others, often make statements containing absolute and unsubstantiated facts. In reality, the properties that concern geographers are all relative. A town, a factory, a farm, or other feature is absolute, but the interest or relevance of this feature to geographers lies in one or more of the properties of the town, factory, etc.—its location, distance from *x*, production, population, size, etc. If the feature in question is not being compared with another tangible feature, it is related to some hypothetical or ideal situation. For instance, it is impossible to describe something as being 'big' without subconsciously relating it to something else.

Many of the assertions made by geographers, then, are not absolutely true, but are more often likely or unlikely or can be seen as partially true. Examples would include such statements as: 'the highlands are devoid of settlement'; 'the boot and shoe industry is footloose'; 'the main farming area is in the valley'; 'the climate is too cold for the cultivation of the vine'. Heisenberg's principle of uncertainty states that all observations of the natural world contain this essential uncertainty. If we believe, then, that we live in 'a world which is neither wholly rational nor wholly chaotic, but is a probabilistic amalgam of choice, calculation and chance' (P. Haggett, *Locational Analysis in Human Geography*, Longman), we should accept or reject these statements in terms of the likelihood of their truth. This likelihood of truth can be seen as a continuum from the absolute falsehood to the absolute truth, as shown in Fig. 18. This shows that our statements can be viewed in terms of probability theory, and it is necessary to use this idea to improve the meaning and objectivity of what is asserted.

Fig. 18. The scale of probability

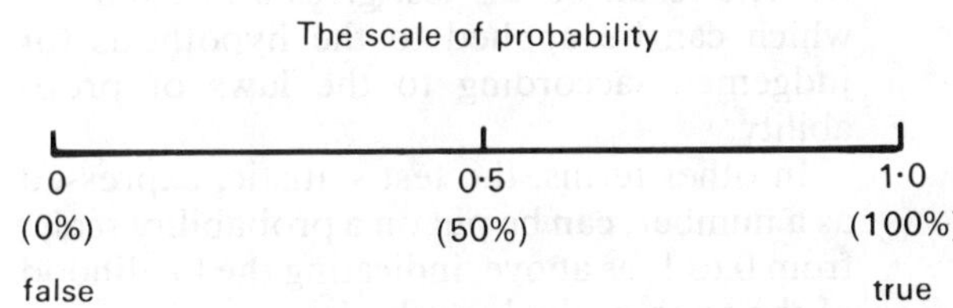

A well-defined procedure helps the geographer to assess this likelihood of truth in a particular assertion or proposition:

(a) Firstly, an assertion or statement about the map under study or the real-life situation is made. For example: 'settlements in this region are to be found in the valley.'

(b) This assertion is then re-stated as a statistical hypothesis—i.e., it is stated in such a way that a precise inference can be made from it. For example: 'the number of settlements above a given size found in the valley is significantly higher than the number of settlements found elsewhere in the region.'

(c) Conventionally this is stated as a null hypothesis—i.e., an hypothesis of no difference between the observed (real life) situation and that situation to be expected by pure chance. The null hypothesis is generally denoted as H_0. In the example above, H_0 might be 'there is no significant difference between the number of settlements above a given size in the valley and the number of settlements outside the valley'.

(d) Evidence from the map or region is then collected to test this hypothesis. For the example above the evidence may be shown on the following lines:

Table 14 Data collected to test hypothesis

	Settlements of more than 5,000 pop.	*Settlements of less than 5,000 pop.*
Valley location		
Non-valley location		

(e) This evidence is tested to determine whether the actual (or observed) distribution of settlement is significantly higher than the distribution to be expected according to chance.

(f) The result of this test gives a test statistic which can be applied to the hypothesis for judgement according to the laws of probability.

In other terms, the test statistic, expressed as a number, can be put on a probability scale, from 0 to 1, as above, indicating the likelihood of the truth of the hypothesis.

It is very important to note, however, that the number itself (the test statistic) does not determine whether we accept or reject the hypothesis. The level of probability to be reached by the test statistic before it can be inferred that the result is significant is set by the person using the test. This level is determined according to the accepted levels of statistical significance, which allow the researcher/tester to impose differing degrees of rigour on the result of the test statistic. (See chapter 8.)

The full meaning of this kind of test is best seen in an application of the Chi-squared test.

The Chi-squared test. The Chi-squared test (giving the test statistic χ^2) is used to find the significance of the difference between one or more frequency distributions and an hypothesised or ideal or expected distribution. The test can only be used if the data is in the form of classes, groups or frequencies. Normally, then, it is applied to data in contingency tables. (See Table 14.)

Fig. 19, for example, shows three groups of field workers studying three different points, A, B and C, along the same stream. Each group takes a sample of 50 pebbles (selected randomly) and measures their length.

Fig. 19 Three sampling points along a stream

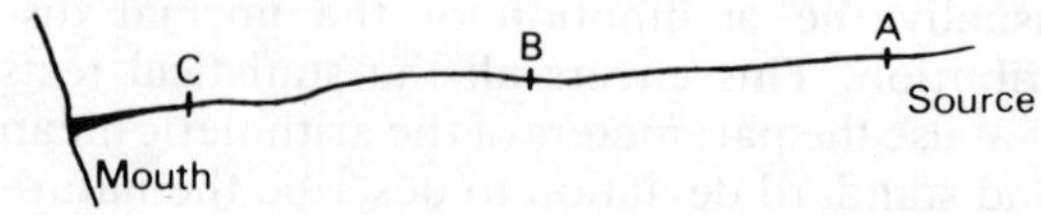

Between them they are trying to test the hypothesis that: 'the pebbles in the river bed become progressively smaller from source to mouth.'

The data collected by the three field groups is shown in Table 15, and the aim is to compare the frequencies of pebble occurrence in each size class with the frequencies that could be expected to occur by chance.

Table 15 Pebble size at three river sites

Size class (mm) \ *Sites*	A	B	C	*Total*
41 mm and over	20	5	2	27
21–40 mm	25	33	12	70
1–20 mm	5	12	36	53
Total	50	50	50	150

Having made an assertion about the real situation and having stated this as an hypothesis, it is now necessary to re-state this as a null hypothesis. This is a statement of no difference, and might be: 'there is no significant difference in the frequency of occurrence between sites of pebbles of different sizes other than might be expected by chance.' If the first assertion made is correct, then the null hypothesis will be rejected.

The formula for the test is given as:

$$\chi^2 = \Sigma \frac{(O - E)^2}{E}$$

where:

O = observed (actual) pebble size frequencies

E = expected pebble size frequencies

The observed pebble size frequencies are those found by the three groups of field workers. The expected pebble size frequencies are calculated by:

$$E = \frac{\text{row total . column total}}{\text{grand total}}$$

For example, in the case of the largest size class of pebbles at site A, the expected value would be:

$$E = \frac{27\ .\ 50}{150}$$

$$\therefore E = 9$$

That is to say that, given a total of 50 pebbles at each site out of a grand total of 150—i.e., a third—if a total of 27 were counted in this size class, then a third of these could be expected to occur at each site, by chance.

For each box in the matrix or contingency table (Table 15) the expected value is calculated in the same way. The observed and expected values are shown together in Table 16. This shows that for the pebble-size class 41 mm and over at site A there are 11 more pebbles observed than would be expected by chance (i.e., 20 − 9 = 11).

Table 16 Observed and expected values of pebble size at three river sites

Size class (mm) / Sites	A	B	C	Total
41 mm and over	20 **9**	5 **9**	2 **9**	27
21–40 mm	25 **23·3**	33 **23·3**	12 **23·3**	70
1–20 mm	5 **17·7**	12 **17·7**	36 **17·7**	53
Total	50	50	50	150

(Expected values in bold type)

As the formula suggests, for each size class at each site, the amount by which O differs from E is squared and related to E. The sum of these values gives the total sum of the squares of the differences between the observed and expected values. This is χ^2.

For the example shown in Tables 15 and 16 the calculations are as follows:

$$\chi^2 = \Sigma\frac{(O-E)^2}{E}$$

$$= \Sigma\left\{\frac{(20-9)^2}{9} + \frac{(5-9)^2}{9} + \frac{(2-9)^2}{9} + \frac{(25-23{\cdot}3)^2}{23{\cdot}3} + \frac{(33-23{\cdot}3)^2}{23{\cdot}3} + \frac{(12-23{\cdot}3)^2}{23{\cdot}3} + \frac{(5-17{\cdot}7)^2}{17{\cdot}7} + \frac{(12-17{\cdot}7)^2}{17{\cdot}7} + \frac{(36-17{\cdot}7)^2}{17{\cdot}7}\right\}$$

$$= \Sigma\left\{\frac{11^2}{9} + \frac{-4^2}{9} + \frac{-7^2}{9} + \frac{1{\cdot}7^2}{23{\cdot}3} + \frac{9{\cdot}7^2}{23{\cdot}3} + \frac{-11{\cdot}3^2}{23{\cdot}3} + \frac{-12{\cdot}7^2}{17{\cdot}7} + \frac{-5{\cdot}7^2}{17{\cdot}7} + \frac{18{\cdot}3^2}{17{\cdot}7}\right\}$$

$$= \Sigma\left\{\frac{121}{9} + \frac{16}{9} + \frac{49}{9} + \frac{2{\cdot}9}{23{\cdot}3} + \frac{94{\cdot}1}{23{\cdot}3} + \frac{127{\cdot}7}{23{\cdot}3} + \frac{161{\cdot}3}{17{\cdot}7} + \frac{32{\cdot}5}{17{\cdot}7} + \frac{334{\cdot}9}{17{\cdot}7}\right\}$$

$$= \Sigma\{13{\cdot}4 + 1{\cdot}8 + 5{\cdot}4 + 0{\cdot}1 + 4{\cdot}0 + 5{\cdot}5 + 9{\cdot}1 + 1{\cdot}8 + 18{\cdot}9\}$$

$$\therefore \chi^2 = 60{\cdot}0$$

Before establishing the level of significance of this χ^2 result, it is necessary to determine the degrees of freedom that constrain the statistic. As mentioned in previous chapters, the degrees of freedom give some meaning to the test statistic, in this case χ^2, by relating it to the number of items in the sample upon which the test is based. Here, the test is based upon values in three rows and three columns, and the degrees of freedom are given by:

$$df = (r-1)\ .\ (c-1)$$

where:

r = the number of rows

c = the number of columns

In the example given above:

$$df = (3 - 1) \,.\, (3 - 1)$$
$$= 2 \,.\, 2$$
$$\therefore\ df = 4$$

The value of χ^2 and the value of the degrees of freedom are then referred to the graph shown in Fig. 20 (or to Table 5 of Lindley and Miller: *Cambridge Elementary Statistical Tables*). If the value falls above the 0·1% line, this means that there is less than 0·1% probability that the null hypothesis is true. So there is greater than 99·9% probability that the alternative hypothesis, denoted as H_1, is true.

H_0 is therefore rejected and it can be concluded that statistically there is a highly significant difference in pebble sizes at the three different sites along the river.

This difference in pebble size could occur by chance on only one occasion in every 1,000.

Fig. 20 Chi-squared significance graph

(Based on data in Cambridge Elementary Statistical Tables and reproduced here from *Statistical Methods and the Geographer*, by S. Gregory (Longman, 1963).)

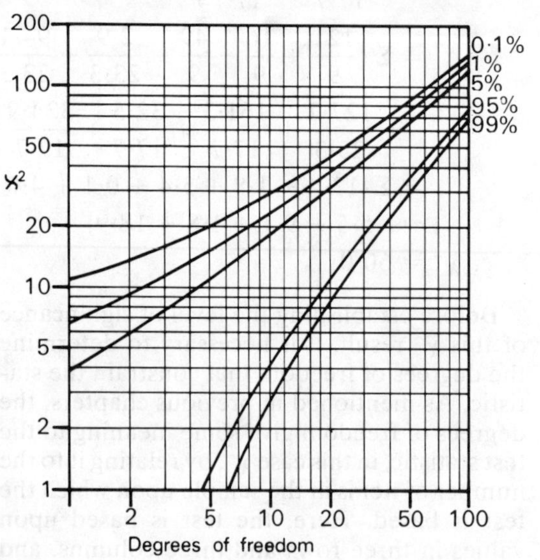

Some factor other than chance, then, must be at work producing pebbles of different size at each site. What might this factor be? This question illustrates the very simple point that thought is still necessary when statistical techniques are used. Statistics are no substitute for intelligence, but they can help the researcher, discursant or student to reach something other than arbitrary conclusions.

Summary of stages in the application of χ^2.

1 Assertion made.

2 This assertion is converted into a null hypothesis, often in the form of: 'there is no significant difference between the frequency of . . .'.

3 The data is collected, arranged in classes or groups and put in matrix form. This matrix holds the observed frequencies.

4 The totals are calculated.

5 The frequencies expected by chance are calculated for each box in the matrix by:

$$E = \frac{(\text{row total})\,.\,(\text{column total})}{\text{grand total}}$$

6 Apply the test statistic, χ^2:

$$\chi^2 = \Sigma \frac{(O - E)^2}{E}$$

7 Calculate the degrees of freedom. Where the matrix holds more than one row and more than one column, these are given by:

$$df = (r - 1)\,.\,(c - 1)$$

Where the data involves only one independent sample—i.e., there is only one row or only one column in the matrix—then they are given by:

$$df = (n - 1)$$

8 The level of significance of the result is read from the appropriate graph or table.

9 The null hypothesis is either accepted or rejected, depending on the requisite level of significance—this to be set by the student before applying the test.

10 An inference is made from the result.

11 The student collapses with mental exhaustion.

11 Exercises

1 A fertilizer is claimed to increase yields. In an experiment on 164 farms, half of them used the fertilizer and a half did not. The resulting yields are shown in Table 17. Test the null hypothesis that there is no relationship between fertilizer application and yields.

Table 17 The results of fertilizer application on 164 farms

	Increased yield	Decreased yield	No change	Total
Fertilizer applied	52	10	20	82
Fertilizer not applied	44	12	26	82
Total	96	22	46	164

$$df = (r-1)(c-1) = (2-1)(3-1) = 2$$

2 Fig. 21 shows the survival of a plant species in a tidal estuary.
(a) Show the data from the map in a 2 by 2 matrix or contingency table.
(b) Formulate a null hypothesis to test whether there is evidence to suggest that the chance of survival is less on the higher ground.

3 Three major areas were defined in a town, and a building survey in each area classified buildings according to their degree of decay. The information collected is shown in Table 18.

Fig. 21 The survival of a plant species in a tidal estuary

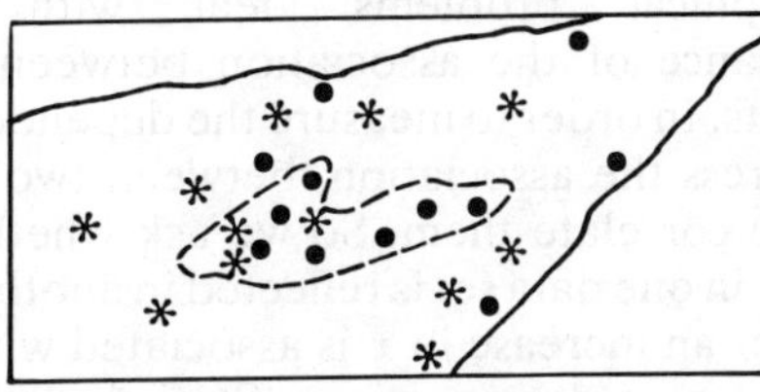

(a) Test the null hypothesis that there is no significant difference in the degree of building decay between the three areas.
(b) How specific are the inferences that can be drawn from this result?
(c) How could the data in Table 18 be re-grouped to give a different result from which different or more specific inferences could be made?

Table 18 Degree of building decay in three areas

	Degree of decay				
Area	1 (Lowest)	2	3	4 (Highest)	Total
1	69	30	12	9	120
2	47	32	5	3	87
3	57	19	17	10	103
Total	173	81	34	22	310

12 Correlation

We have covered the statistical significance of the difference between two data sets, but many geographical problems deal with the significance of the association between two data sets. In order to measure the dependence, or express the association, between two data sets we correlate them. So we ask whether a change in one data set is reflected in another or whether an increase in x is associated with an increase or a decrease in y. Statistically correlating two data sets gives us a correlation coefficient.

The product moment correlation coefficient. Table 19 shows the population per hospital bed and the male life expectancy of selected countries. Let us assume that we wish to assess the statistical significance of the relationship between these two. In other words, we are asking whether an increase in life expectancy is associated with a decrease or increase in the population per hospital bed.

Let the life expectancy data be denoted by x and the population per hospital bed be denoted by y. We first take each value of x and compare it with its arithmetic mean, $\bar{x}$, and do the same for y. We show not only the difference between x and $\bar{x}$ but also the direction of difference, positive or negative. Each pair of values is then multiplied together to give the product. As shown in the fifth and sixth columns of Table 20, this product will be either positive or negative. These two columns are then totalled to give the sum of the product of differences from the mean:

$$\Sigma(x - \bar{x}) \cdot (y - \bar{y}).$$

It is necessary to relate this measure to the number of items in the sample. So the above sum is divided by n, where n = the number of pairs in the sample data sets, to give what is known as the *covariance*. This is given by:

$$\frac{\Sigma(x - \bar{x}) \cdot (y - \bar{y})}{n}$$

Having rendered the measure independent of the number in the sample by dividing by n,

Table 19 Life expectancy and population per hospital bed

Country	Male life expectancy at birth (years) (average 1960/5) x	Population per hospital bed (1965) y
Argentina	63·7	160
Australia	67·9	80
Barbados	62·7	150
Belgium	67·7	130
Costa Rica	61·9	250
Denmark	70·1	110
Dominican Rep.	57·2	350
El Salvador	56·6	460
West Germany	67·6	90
Greece	67·5	170
Guyana	59·0	170
Italy	67·2	100
Jamaica	62·6	260
Mexico	57·6	510
Netherlands	71·1	200
New Zealand	68·4	100
Peru	52·6	410
Portugal	60·7	170
Sweden	71·6	70
Switzerland	68·7	80
Trinidad and Tobaga	62·2	230
Venezuela	66·4	310
	Σ_x = 1,411·0 n = 22 $\bar{x}$ = 64·1 σ_x = ±5·12	Σ_y = 4,560 n = 22 $\bar{y}$ = 207 σ_y = ±125·4

(Compiled from the *Oxford Economic Atlas of the World*, 1974)

we should now render it independent of the units of measurement by dividing the covariance by the product of the two standard deviations, σ_x and σ_y. This gives the correlation coefficient:

$$r = \frac{\dfrac{\Sigma(x - \bar{x}) \cdot (y - \bar{y})}{n}}{\sigma_x \cdot \sigma_y}$$

Table 20 Calculation of the product moment correlation coefficient—Life expectancy and population per hospital bed

Original data		***Difference from the mean***		***The product*** $(x-\bar{x}) \cdot (y-\bar{y})$	
x	y	$(x-\bar{x})$	$(y-\bar{y})$	***positive***	***negative***
63·7	160	−0·4	−47	+18·8	
67·9	80	+3·8	−127		−482·6
62·7	150	−1·4	−57	+79·8	
67·7	130	+3·6	−77		−277·2
61·9	250	−2·2	+43		−94·6
70·1	110	+6·0	−97		−582·0
57·2	350	−6·9	+143		−986·7
56·6	460	−7·5	+253		−1,897·5
67·6	90	+3·5	−117		−409·5
67·5	170	+3·4	−37		−125·8
59·0	170	−5·1	−37	+188·7	
67·2	100	+3·1	−107		−331·7
62·6	260	−1·5	+53		−79·5
57·6	510	−6·5	+303		−1,969·5
71·1	200	+7·0	−7		−49·0
68·4	100	+4·3	−107		−460·1
52·6	410	−11·5	+203		−2,334·5
60·7	170	−3·4	−37	+125·8	
71·6	70	+7·5	−137		−1,027·5
68·7	80	+4·6	−127		−584·2
62·2	230	−1·9	+23		−43·7
66·4	310	+2·3	+103	+236·9	
				$\Sigma+ = +650·0$	$\Sigma- = -11,735·6$
				$\therefore \Sigma(x-\bar{x})(y-\bar{y}) = -11,085·6$	

In the example tabulated above:

$$\text{the covariance} = \frac{-11,085·6}{22} = -503·9$$

$$\therefore r = \frac{-503·9}{5·12 \,.\, 125·4} = \frac{-503·9}{642·05}$$

$$\therefore r = -0·78$$

The results obtained for r vary vetween $+1$ and -1, where:

$+1$ represents perfect positive or direct statistical correlation;

-1 represents perfect negative or inverse statistical correlation;

0 represents no statistical correlation.

It is very important to recognize that the absence of statistical correlation does not preclude some form of association. It merely indicates that the two data sets display no statistical link as they stand. By the same token, a positive or negative value of r does not necessarily imply a causal relationship between the two phenomena. It simply implies some measure of statistical association between the data sets.

In the example given above this means that we cannot say that a high life expectancy is caused by a low rate of population per hospital bed. There are clearly many other factors involved. But we can say that there is a statistical association between life expectancy and the rate of population per hospital bed.

This point must be stressed and the reader should fully understand that there is a large step between a statistical correlation and a conclusion that an increase in x *causes* an increase or decrease in y. It is also possible to arrive at a nonsense correlation, where two sets of data show a statistical correlation, although it is perfectly clear, by commonsense, that they are not related in any way. Examples of this might include the relationship between the variations in a nation's GNP rate (Gross National Product) over a number of years and the variations in the same nation's penguin population over the same period of time, or, more seriously, regional variations in motorway length and unemployment. With the latter example it may be possible to define an association and to demonstrate a statistical correlation between the two, however tenuous, but it would be folly to suggest that the incidence of one *causes* or *leads to* an increase or decrease in the other.

It is essential, then, to understand and to be wary of these limitations to the technique of statistical correlation. The careful consideration of the appropriateness of the technique and a sensible interpretation of the result obtained are more important than the technique itself.

Back to the calculations. Thus far we know the meaning of extreme results, but extreme values of r are rarely if ever obtained. As a general rule of thumb, values of more than $+0{\cdot}5$ or less than $-0{\cdot}5$ are said to show a relationship that is probably significant, but a further test is required to determine what level of statistical significance is reached. This is done by assessing the probability of obtaining the given value of the coefficient, r, purely by chance, given the two data sets under study. This is achieved by expressing the correlation coefficient in terms of the Student's t statistic and by applying the value for t so derived, as shown in chapter 8. The expression of r in terms of t is given by:

$$t = \frac{r \, . \, \sqrt{n-2}}{\sqrt{1-r^2}}$$

where r = the product moment correlation coefficient and n = the number of pairs of values involved in the correlation.

In order to test the result of t, the graph of the Student's t distribution (Fig. 15) is used, and for this the degrees of freedom must be calculated. These are given by $(n - 2)$—minus two because there are two data sets. The value of t and the degrees of freedom can then be referred to the graph to determine its statistical significance.

For the example already used in this chapter, the calculations would be as follows:

$$t = \frac{-0{\cdot}78 \, . \, \sqrt{22-2}}{\sqrt{1-(-0{\cdot}78)^2}}$$

$$= \frac{-0{\cdot}78 \, . \, \sqrt{20}}{\sqrt{1-0{\cdot}61}}$$

$$= \frac{-0{\cdot}78 \, . \, 4{\cdot}47}{\sqrt{0{\cdot}39}}$$

$$= \frac{-3{\cdot}49}{0{\cdot}62}$$

$$\therefore \; t = -5{\cdot}63$$

$$df = (n-2) = (22-2) = 20$$

In practice the direction of t is ignored for use with the graph as it is only a result of the direction of r, the correlation coefficient, which, being negative, just indicates an inverse relationship.

Reference to the Student's t graph (Fig. 15) shows that this value of t falls above the $0{\cdot}1\%$ significance line. This means that there is less than a $0{\cdot}1\%$ probability (or one in a thousand) that the association shown by this value of r occurred purely by chance. In other terms there is a $99{\cdot}9\%$ probability that there is a true or real association between the two data sets, or it is a result that is 'highly significant'.

This calculation of t can be short-cut by referring the value of r directly to a specially adapted graph of the t distribution. Such a graph is shown in Fig. 22, from which the significance of the value of r can be determined. It can be seen that with 20 degrees of freedom the value of $-0{\cdot}78$ again falls above the $0{\cdot}1\%$ probability line, indicating a 'highly significant' statistical correlation.

The product moment correlation coefficient takes a great deal of time and effort in its calculation, particularly when there are many numbers in each data set. This is of course no problem if a computer is available, but few school geography departments have such facilities. There are however other correlation techniques which employ a ranking procedure rather than the actual values of the items in each data set. These techniques have the advantage that they are easier and less time-consuming in their calculation, but they give slightly less reliable results in that they ignore the actual values of each item in the data set.

Fig. 22 Graph of significance levels for correlation coefficients using Student's t distribution
(From Gregory, S. *Statistical Methods and the Geographer*, Longman, 1963).

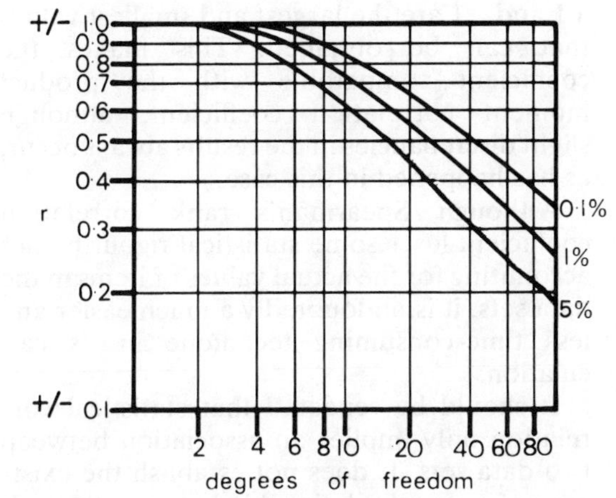

One example of these rank correlation techniques will be outlined here.

Spearman's rank correlation coefficient, or rho. This is sometimes referred to as Spearman's rho, and is variously denoted as V_r, r_s, rho, ρ (Greek letter), or *R*.

The limits of r_s are again +1 and −1, and the graph of significance levels (Fig. 22) can be used to test the value of the coefficient.

Using the same two data sets—i.e., those shown in Table 19—rank 1 is assigned to the highest value, rank 2 to the next highest, rank 3 to the next and so on. This is done for each data set as shown in Table 21. It will be seen from this that some of the items share the same rank. These are known as tied ranks and are assigned the median value of the two or more tied ranks. So, if two countries share the same rank 2—i.e., are both second highest in rank order—then they will both be assigned the rank value 2·5 and the next highest will be rank 4. After putting both data sets into rank order, the difference in rank, *d*, between each pair of values is taken and then squared. The

Table 21 Calculation of Spearman's rank correlation coefficient—male life expectancy and population per hospital bed

Original data		*Rank*		*Difference in ranks, d*	d^2
x	*y*	*x*	*y*		
63·7	160	12	13	1	1
67·9	80	6	20·5	14·5	210·25
62·7	150	13	14	1	1
67·7	130	7	15	8	64
61·9	250	16	7	9	81
70·1	110	3	16	13	169
57·2	350	20	4	16	256
56·6	460	21	2	19	361
67·6	90	8	19	11	121
67·5	170	9	11	2	4
59·0	170	18	11	7	49
67·2	100	10	17·5	7·5	56·25
62·6	260	14	6	8	64
57·6	510	19	1	18	324
71·1	200	2	9	7	49
68·4	100	5	17·5	12·5	156·25
52·6	410	22	3	19	361
60·7	170	17	11	6	36
71·6	70	1	22	21	441
68·7	80	4	20·5	16·5	272·25
62·2	230	15	8	7	49
66·4	310	11	5	6	36
					Σd^2 = 3,162·0

squares of the differences in rank are then summed, and it is this, Σd^2, which forms the basis of the formula for obtaining the test statistic, Spearman's rho, which is given as:

$$r_s = 1 - \frac{6 \,.\, \Sigma d^2}{n^3 - n}$$

In the example given here, the calculations are as follows:

$$r_s = 1 - \frac{6 \,.\, 3{,}162}{22^3 - 22}$$

$$r_s = 1 - \frac{18{,}972}{10{,}626}$$

$$r_s = 1 - 1{\cdot}79$$

$$r_s = -0{\cdot}79$$

With 20 degrees of freedom (calculated as for the product moment correlation coefficient) Fig. 22 shows that this value of r_s demonstrates a 'highly significant' statistical association between the two data sets. That is, the value of −0·79 falls above the 0·1% line, meaning that there is less than a 0·1% probability that this measure of association occurred purely by chance.

The formula for r_s is designed to ensure that +1 and −1 are the largest and smallest values that can be obtained. This makes the coefficient comparable with the product moment correlation coefficient, although slight discrepancies in the results always occur, as has happened in this case.

Although Spearman's rank correlation coefficient loses some statistical rigour by not accounting for the actual values of items in the data sets, it is undoubtedly a much easier and less time-consuming technique in its calculation.

It should be repeated that statistical correlation only implies an association between two data sets. It does not establish the existence of a causal relationship between them. It must also be emphasized that techniques of statistical correlation are not intended as substitutes for thought, but simply as aids to the geographer.

13 Exercises

Table 22 Increase in the population and number of shops in selected new towns

New town	*Population increase since designation to 1970* *x*	*Number of shops since new town designation to 1970* *y*	*Summary data (Arithmetic mean and standard deviation rounded off to whole figures)*
Basildon	52,200	320	Σx = 531,600
Bracknell	27,900	135	
Crawley	57,400	286	n = 15
Harlow	71,000	321	
Hatfield	16,500	103	$\bar{x}$ = 35,440
Hemel Hempstead	47,800	318	
Stevenage	55,000	317	σ_x = ±20,098
Welwyn G.C.	24,000	133	
Corby	32,400	204	Σ_y = 2,668
Telford	700	3	
Cwmbran	32,100	199	n = 15
Cumbernauld	23,500	47	
East Kilbride	60,100	180	$\bar{y}$ = 178
Glenrothes	24,900	72	
Livingstone	6,100	30	σ_y = ± 110

1 Following the procedure demonstrated in the previous chapter, calculate the product moment correlation coefficient that describes the association between the two data sets shown in Table 22.

2 Calculate t and the degrees of freedom for the coefficient found in point 1 above, and refer the results to Fig. 15. With careful and thoughtful use of words draw conclusions from this.

3 Can the same conclusions be drawn, without the calculation of t, using Fig. 22?

4 Assign a rank to each item in the data sets. (Rank 1 to the highest.)

5 Calculate Spearman's rank correlation coefficient for the data in Table 22.

6 Is there a difference in the values of the two coefficients calculated in this chapter? If there is a difference, does it lead to different conclusions from each coefficient? (Answers should again be worded very carefully.)

14 Regression

Very often it is necessary to ascertain not just the extent or strength of a relationship (as shown by the correlation coefficient) but also the form of the relationship as shown by the scatter of points on a graph. This is particularly important in forecasting by the projection of a trend line into the future—see chapter 18.

Fig. 23 shows a graph on which is plotted the scatter of items from the data sets of male life expectancy and population per hospital bed used in chapter 12. This simply shows the scatter of values for the 22 countries. In order to define a trend or pattern from this it is possible to fit a line through the scatter of points to represent the 'average trend'. In other words a line is drawn which describes the average relationship between the two sets of data. This can be done by pure guesswork, as with line AA in Fig. 23. Alternatively, it is possible to express such a linear relationship more accurately by calculating mathematically the exact position of this 'average' line. When calculated mathematically these lines are known as *regression lines*, and they describe the 'best fit' line through a series of points on a graph.

Fig. 23 Male life expectancy and population per hospital bed—selected countries

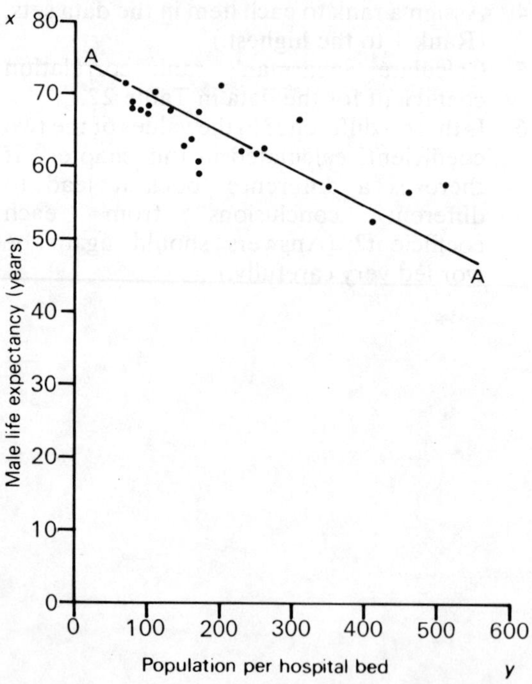

Straight line regression. In the case of perfect correlation between two data sets, all points shown on the graph would fall on the regression line. Fig. 24(a) shows the regression line, BB, for two data sets which are correlated perfectly in a positive way, such that when x increases, y increases. Fig. 24(b) shows the regression line obtained when there is perfect negative correlation between the data sets.

Fig. 24 Two ideal regression lines

(a) Perfect positive correlation

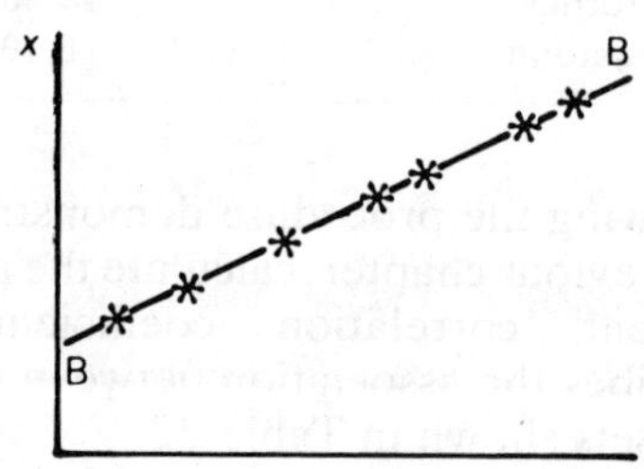

(b) Perfect negative correlation

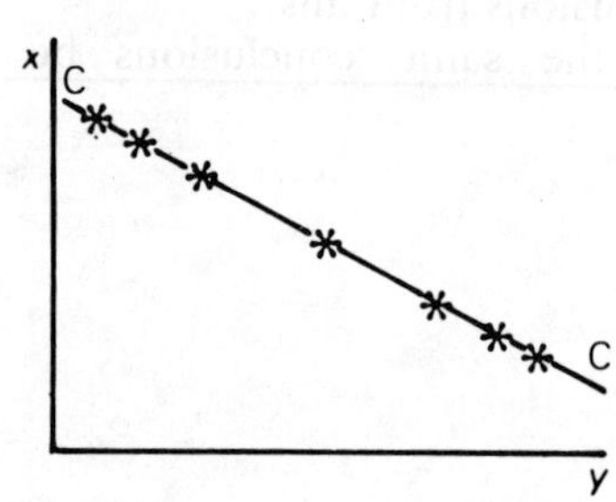

Of course, in practice, perfect correlation is rarely if ever obtained. The relationship is likely to be less regular and clear-cut, as it is in Fig. 23.

In obtaining the regression line by calculation, the aim is to ensure that the sum of the squares of the differences (or deviations) of the individual values from the line is at an absolute minimum. Fig. 25 shows that the line takes that course which minimizes the sum of the squares of all values of *a*. It is known as the method of least squares.

Fig. 25 Obtaining the regression line

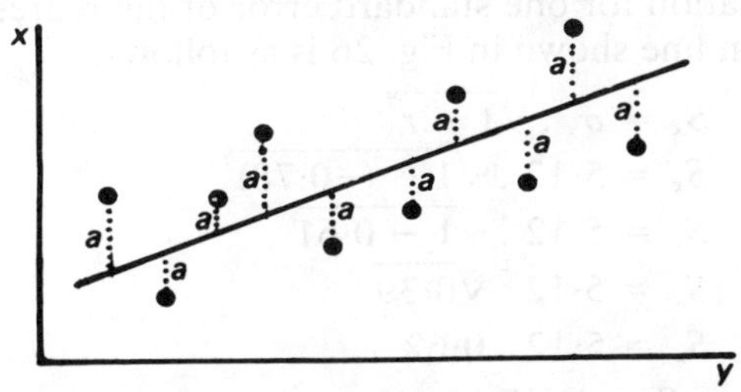

(a = deviations from the line of individual items)

It is theoretically possible to calculate the minimum value for the sum of squares by setting up the appropriate equation for each pair of values under consideration. But this is very lengthy, so a formula is applied to give the same results with a lot less labour. This requires the correlation coefficient, r, the arithmetic means, $\bar{x}$ and $\bar{y}$, and standard deviations, σ_x and σ_y. It is given by:

$$x - \bar{x} = r \cdot \frac{\sigma_x}{\sigma_y} \cdot (y - \bar{y})$$

where: x is unknown
y is given (known)

So the unknown value, x, differs from the arithmetic mean of its data set, $\bar{x}$, by the same amount as the known value, y, differs from its arithmetic mean, $\bar{y}$, modified by the ratio of the two standard deviations, which expresses the overall spread of values about their respective averages, and by the correlation coefficient, r, which expresses the degree of relationship between the two data sets. In the example used here, the regression equation is found in the following way:

$$x - 64{\cdot}1 = -0{\cdot}78 \cdot \frac{5{\cdot}12}{125{\cdot}4} \cdot (y - 207)$$

$$x - 64{\cdot}1 = -0{\cdot}78 \,.\, 0{\cdot}041 \,.\, (y - 207)$$

$$x - 64{\cdot}1 = -0{\cdot}032(y - 207)$$

$$x - 64{\cdot}1 = -0{\cdot}032y + 6{\cdot}6$$

$$\therefore x = -0{\cdot}032y + 6{\cdot}6 + 64{\cdot}1$$

$$x = -0{\cdot}032y + 70{\cdot}7$$

This is the regression equation, and with this it is possible to calculate the value of x for any given value of y. So, assume that:

$$y = 0,$$

then:

$$x = (0 \,.\, -0{\cdot}032) + 70{\cdot}7$$

$$\therefore x = 70{\cdot}7$$

In the above equation $-0{\cdot}032$ is known as the *regression coefficient*, and gives the slope of the regression line, whilst 70·7 is the *base constant* or *intercept value*, and indicates the point of origin of the line—i.e., the value of the unknown variable, x, when the known or given variable, y, is zero.

We now have one pair of co-ordinates for the regression line. As the line is straight, only one more pair of co-ordinates is required. Assume that $y = 500$, then:

$$x = (500 \,.\, -0{\cdot}032) + 70{\cdot}7$$

$$x = -16 + 70{\cdot}7$$

$$\therefore x = 54{\cdot}7$$

The two pairs of co-ordinates for the regression line, then, are:

$$y = 0; \quad x = 70{\cdot}7$$

$$y = 500; \quad x = 54{\cdot}7$$

These points can now be plotted on the graph and joined to give the regression line, as shown in Fig. 26, which illustrates the closest approximation to the relationship between the two data sets.

With this regression line it is possible to predict the expected value for one data set given a value for the other. Such a value is, however, subject to error as the regression line gives only the closest approximation to the relationship. It is possible, though, to calculate the standard error of this 'best-fit' line so that the probability of the prediction being correct can be given. (The concept of the standard error was first introduced in chapter 8.)

Fig. 26 The regression of male life expectancy against population per hospital bed

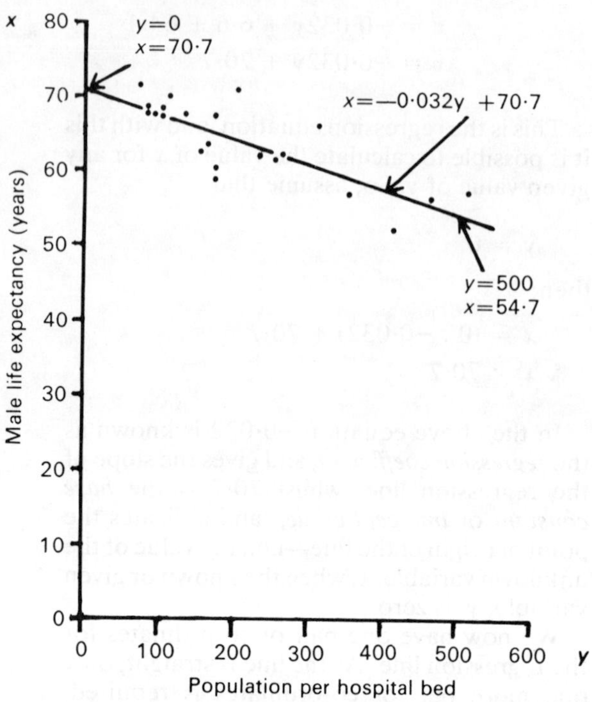

Standard errors and confidence limits. In anything but the perfect relationship it can be said that the regression line is only a 'best-fit' line. It is therefore often desirable to calculate the standard error of this 'best-fit' line in order to determine the probability of being correct in the prediction made from it. Thus, the range within which actual conditions are likely to fall can be assessed with some accuracy.

The standard error of the predicted or expected unknown value, x, is expressed by the term S_x, and is calculated by the formula:

$$S_x = \sigma_x \sqrt{1 - r^2}$$

When this value is determined it can be said that there is a 68% probability that actual values will differ from the regression line by not more than the standard error, and that there is a 95% probability that actual values will differ from the regression line by not more than two times this standard error. Put in another way, this means that the probability of values differing from the regression line by more than two times the standard error is only 5% at the most. The 99% level of probability can also be applied to the regression line at three times its standard error.

These probability levels applied to the regression of one data set upon another are often referred to as confidence limits, and can be drawn parallel to the regression line on the graph. As they are parallel, it is only necessary to calculate one point for each standard error or confidence limit.

Using this chapter's example, the calculation for one standard error of the regression line shown in Fig. 26 is as follows:

$$S_x = \sigma_x \sqrt{1 - r^2}$$
$$S_x = 5{\cdot}12 \,.\, \sqrt{1 - (-0{\cdot}78)^2}$$
$$S_x = 5{\cdot}12 \,.\, \sqrt{1 - 0{\cdot}61}$$
$$S_x = 5{\cdot}12 \,.\, \sqrt{0{\cdot}39}$$
$$S_x = 5{\cdot}12 \,.\, 0{\cdot}62$$
$$\therefore S_x = 3{\cdot}17$$

The calculation for two and three standard errors is as follows:

$$2S_x = 2 \,.\, 3{\cdot}17$$
$$\therefore 2S_x = 6{\cdot}34$$
$$3S_x = 3 \,.\, 3{\cdot}17$$
$$\therefore 3S_x = 9{\cdot}51$$

Fig. 27 The regression line and its confidence limits

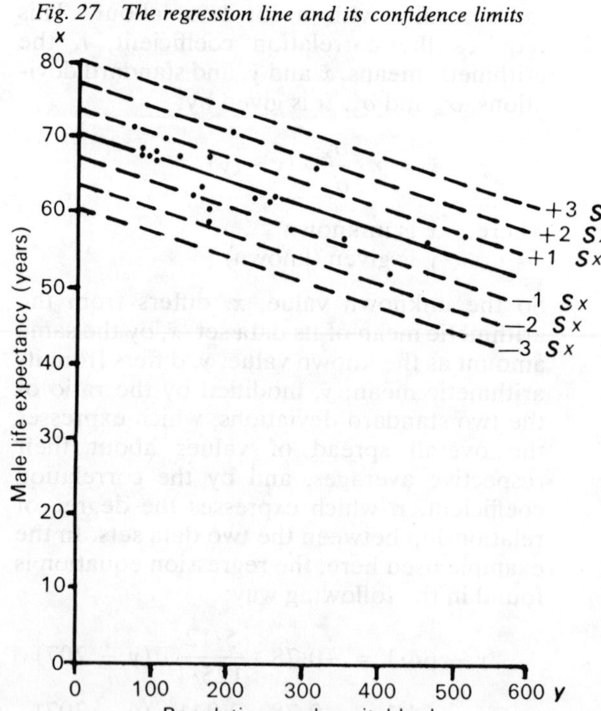

So, when $y = 0$,

$x = 70{\cdot}7 \pm 3{\cdot}17$ with a 68% probability

and $x = 70{\cdot}7 \pm 6{\cdot}34$ with a 95·45% probability

and $x = 70{\cdot}7 \pm 9{\cdot}51$ with a 99·7% probability

So, there is a 95·45% probability that items in the data set of male life expectancy should lie within 6·34 years of the regression line.

These confidence limits for the regression line are shown in Fig. 27. According to probability theory, they express the confidence that can be placed in the use and interpretation of the regression line.

15 Exercises

1 Compare line AA shown in Fig. 23 and the regression line in Fig. 26 for the data sets used in chapters 12 and 14. Comment upon the reliability of both of these lines.

Table 23 Percentage of working population employed in agriculture and the number of tractors per thousand population, 1975, for selected countries

Country	% employed in agriculture x	Number of tractors per '000 population y
Argentina	19·2	8·0
Australia	10·9	26·2
Belgium	5·8	10·6
Brazil	51·6	2·8
Canada	7·9	29·9
El Salvador	60·3	0·8
Ghana	58·0	0·4
Greece	53·9	9·9
Haiti	83·2	0·1
India	72·9	0·4
Israel	12·5	6·8
Jamaica	36·1	1·4
Kenya	35·3	0·5
Mexico	54·2	2·9
Netherlands	10·7	13·8
New Zealand	14·4	34·0
Pakistan	75·0	0·6
Paraguay	52·2	1·1
Peru	49·7	0·9
Sweden	13·8	23·1
USA	6·2	20·2
Venezuela	32·1	2·2
	$\Sigma x = 815{\cdot}9$ $n = 22$ $\bar{x} = 37{\cdot}1$ $\sigma_x = \pm 24{\cdot}47$	$\Sigma y = 196{\cdot}6$ $n = 22$ $\bar{y} = 8{\cdot}9$ $\sigma_y = \pm 10{\cdot}58$
Product moment correlation coefficient $r = -0{\cdot}73$		

2 Table 23 shows the percentage of working population employed in agriculture and the number of tractors per thousand population, 1975, for selected countries.
(a) Construct a graph to show the information given in Table 23.
(b) Simply by looking at the points plotted on your graph draw in a line to represent the 'average relationship' between the two data sets.
(c) Mathematically calculate the regression line that represents the relationship between the two data sets.
(d) On your graph mark in and join up the two points calculated in (c) above.
(e) Comment upon the differences between the two lines drawn in (b) and (d) above.
(f) Calculate points for one and two standard errors of the regression line drawn in (d) above.
(g) Mark in the confidence limits shown by these standard errors on your graph.
(h) State the meaning of the standard errors as shown on your graph.
(i) How closely does the distribution of the sample of points on your graph fit the distribution to be expected according to probability theory?

3 (a) Rank the data shown in Table 23.
(b) Calculate Spearman's rank correlation coefficient for the data shown in Table 23.
(c) i. Does Spearman's rank correlation coefficient differ from the product moment correlation coefficient?
ii. If so, by how much?
iii. Also if so, does this difference lead to results at different levels of statistical significance?
iv. Can you account for this difference?

16 Further measures of areal distributions

Chapter 3 demonstrates ways of measuring the centre of a distribution of features over a given area. Often, however, it is the nature of the whole distribution of objects that concerns geographers rather than just the centre of that distribution. Words such as 'scattered', 'clustered' or 'nucleated' are used to describe geographical distributions. While these descriptive terms convey a simple meaning, they lack precision, are not supported by measurement and allow much room for subjective use and different interpretations.

To counter this problem certain relatively simple statistical techniques have been developed to assist in the description of a distribution or in the measurement of areal concentration. A few of these are outlined below.

Nearest neighbour analysis. This technique can be employed to analyse the distribution of settlements or other features, although it is important that the reader is aware of the fact that there are rather severe limitations on the interpretation of results. The method has been and still is wrongly used, for instance, to measure 'pattern', when in fact it can only yield information on the distribution of distances between points and their nearest neighbours.* Further limitations are outlined later in the text.

It is based upon the measurement of the straight-line distance between a point and its nearest neighbour—i.e., the point nearest to it. This distance is compared with that which might be expected if the points were distributed in a random fashion over the same area.

When the area under study and the nature of the objects being considered have been defined (e.g., the size of settlements), the procedure for the analysis is as follows:

(a) The distance between each point and its nearest neighbour is measured.

(b) These distances are then totalled.

(c) The total is divided by the number of measurements taken, n. This gives $\bar{D}_{obs}$, which is the average of the observed distances from each point to its nearest neighbour.

(d) The density of points in the area under study is then determined by:

$$d = \frac{n}{A}$$

where d = the density
n = the number of points in the area under study
A = the area under study.

(The area should be measured in the same units as the distances. For example, if the distances are measured in kilometres, then the area should be given in square kilometres.)

(e) The expected mean of a random distribution of points over the same area is then calculated by:

$$\bar{r}_e = \frac{1}{2\,.\,\sqrt{d}}$$

where $\bar{r}_e$ = the expected mean of a random distribution ($\bar{r}_e$ is sometimes denoted as $\bar{D}_{ran}$)
d = the density of points

(f) The nearest neighbour statistic can then be obtained by:

$$R_n = \frac{\bar{D}_{obs}}{\bar{r}_e}$$

where

R_n = the nearest neighbour statistic

The values of R_n range from 0 to 2·15. A value of 0 means that all the points are clustered together at one place. The other extreme value, 2·15, indicates that the points are as far as possible from each other, forming a perfectly ordered and uniform distribution, in which all the points lie at the vertices of equilateral triangles. A value of 1·0 suggests a random distribution. These three cases are illustrated in Fig. 28.

* For a more detailed outline see D. Sibley, 'On pattern and dispersion'. *Area*, 1976, Vol. 8, No. 3.

Fig. 28 Three point distributions

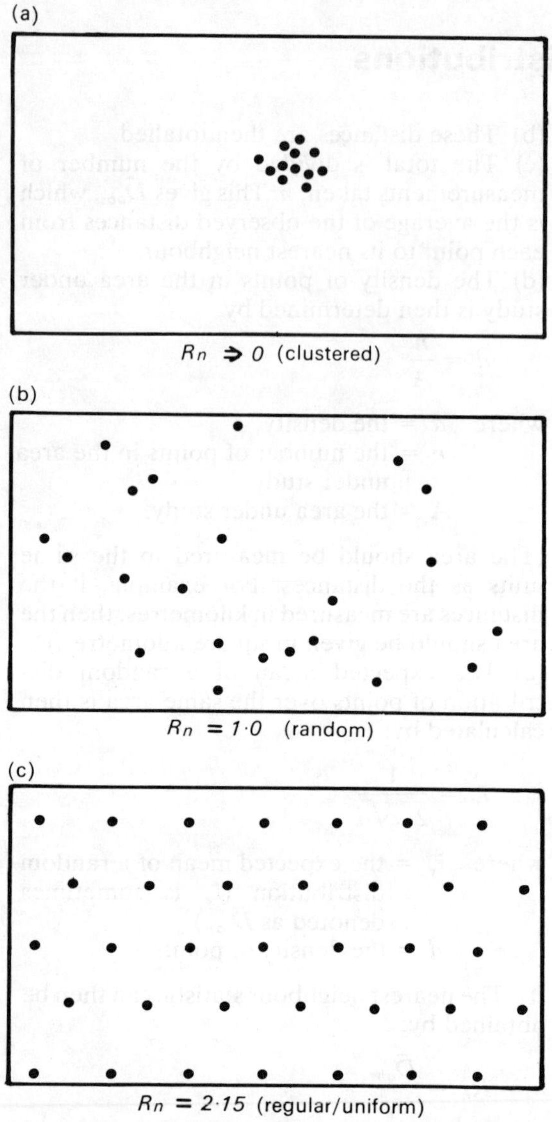

An example should help the reader to follow the technique and to appreciate some of its uses. Fig. 29 shows the point distribution of tumuli (ancient burial mounds) in a part of the Isle of Purbeck, Dorset.

Forty-five tumuli are marked by dots on the map, and the distance from each dot to its nearest neighbour was measured. The total of these distances is 19·2 km. This is divided by the number of dots, 45, to give the average distance to the nearest neighbour—i.e., $\bar{D}_{obs}$.

Fig. 29 The distribution of tumuli in the Isle of Purbeck (Area south of northing 085 and east of easting 391 on O.S. sheet 195)

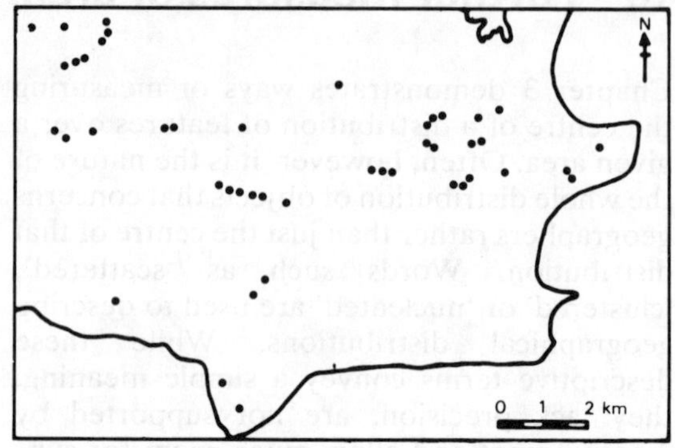

In this case:

$$\bar{D}_{obs} = \frac{19{\cdot}2}{45} = 0{\cdot}43 \text{ km}$$

The area under study covers 100 square kilometres, so the density of points is given by:

$$d = \frac{n}{A} = \frac{45}{100} = 0{\cdot}45$$

When we have obtained this measure of density, it is then a simple matter to calculate the mean distance that would be expected in a random distribution.

$$\bar{r}_e = \frac{1}{2 \,.\, \sqrt{d}}$$

$$\therefore \; \bar{r}_e = \frac{1}{2 \,.\, \sqrt{0{\cdot}45}}$$

$$= \frac{1}{2 \;\; 0{\cdot}67}$$

$$= \frac{1}{1{\cdot}34}$$

$$= 0{\cdot}75$$

So the nearest neighbour statistic can then be obtained.

$$\bar{R}_n = \frac{\bar{D}_{obs}}{\bar{r}_e}$$

$$= \frac{0{\cdot}43}{0{\cdot}75}$$

$$\therefore \; \bar{R}_n = 0{\cdot}57$$

This result suggests an element of clustering, but its level of statistical significance can be checked against a prepared graph to establish a more precise meaning for the statistic and thereby to render the result of more use to the researcher or student. Such a graph is shown in Fig. 30, from which it can be seen that the value of 0·57 in the example used here shows a statistically significant element of clustering in the distribution of tumuli over the given area. Put in precise terms, this means that there is less than a 5% probability that this result would have been obtained with a random distribution.

Fig. 30 0·05 significance level of the R_n statistic

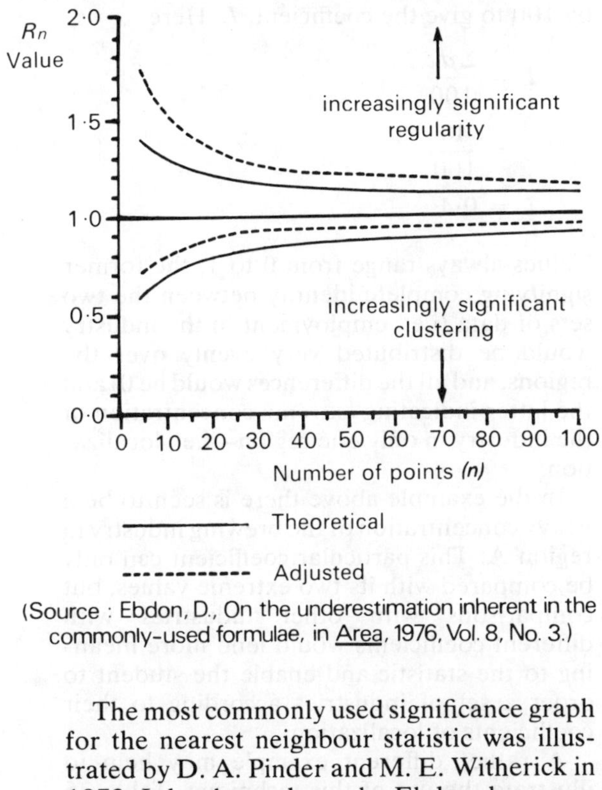

(Source: Ebdon, D., On the underestimation inherent in the commonly-used formulae, in Area, 1976, Vol. 8, No. 3.)

The most commonly used significance graph for the nearest neighbour statistic was illustrated by D. A. Pinder and M. E. Witherick in 1972. It has been shown by Ebdon, however, that the formula for the nearest neighbour statistic contains a consistent underestimation of the true value of the mean nearest neighbour distance. The dashed lines in Fig. 30, then, denote the 0·05 significance level adjusted for this underestimation, and it is these that should be used here. Recently, Dr Pinder has formulated a correction factor for inclusion in the formula so that it allows for this underestimation. This will mean that the theoretical significance lines shown in Fig. 30 can be used. The explanation of and derivation of this correction factor, though, are too lengthy and complicated for inclusion in this book. Readers who wish to follow the technique further should refer to the magazine *Area*, Vol. 10, No. 5, 1978, pp. 379–385.

The nearest neighbour statistic describes the distribution, then, but it is now up to the geographer/student/planner to explain why such a distribution occurs. Clearly, however, the statistic facilitates comparisons between distributions and lends a little more meaning to previously loose terms, such as 'scattered', 'dispersed', 'clustered', 'regular', etc. To date, the uses of this technique have mainly, but not exclusively, concerned settlement distributions and it obviously has relevance to central place theory.

The use of the technique and interpretation of the statistic can be somewhat restricted, though, by any of the following problems:

1 The problem of the area under study has already been outlined in chapter 3 and illustrated in Fig. 11. The same distribution of points may produce differing impressions when considered over different areas and at different scales.

2 Boundary definition also presents a problem, as when the nearest neighbour of a point within the boundary is outside the boundary. Fig. 31 shows that the nearest neighbour to point 1 lies outside the boundary of the area under study. Probably the best solution to this is to include point 2 in the analysis only as a nearest neighbour to point 1. This is not always possible of course and its desirability may depend on the requirements of the study.

Fig. 31 The problem of boundary definition

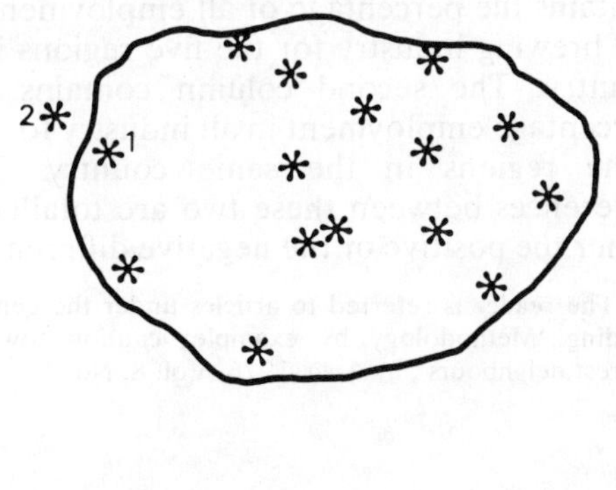

3 It is important to remember that, although a distribution of settlements or of other points may be described by the statistic as random, this does not necessarily mean that the forces producing that distribution have been entirely random.

4 Reliability in the statistic increases as the number of points under consideration increases, and a number of 30 points is often quoted as the minimum for study. The problem with this is that the greater the number of points, the more time-consuming and tedious the measurements.

5 Any number of point distributions can give the same result for R_n. In fact, as mentioned earlier, the statistic is not a measure of pattern, but simply of the distribution of neighbour distances.*

6 It has been shown that the theoretical significance lines given by Fig. 30 produce a consistent bias in favour of regularity.* There are clear dangers in searching for a regular distribution of points with a technique which has a bias towards finding such distributions.

The coefficient of localization. This is sometimes called the index of concentration and it measures the degree of localization or concentration of a given industry or other feature over a set of regions. It compares the percentage distribution of, for example, employment in a particular industry over a number of regions with the percentage distribution of employment in all industry over the same set of regions.

The coefficient, L, is the sum of either the positive or the negative deviations of the regional percentages of workers in the given industry from the corresponding regional percentages of workers in all industry, divided by 100.

This is best illustrated by a simple hypothetical example. The first column in Table 24 contains the percentage of all employment in the brewing industry for the five regions in a country. The second column contains the percentage employment in all industry for the same regions in the same country. The differences between these two are totalled—either the positive or the negative differences,

* The reader is referred to articles under the general heading 'Methodology by example: caution towards nearest neighbours', in *Area*, 1976, Vol. 8, No. 3.

Table 24 The coefficient of localization—a hypothetical example

Region	% of all employment in the brewing industry	% of all employment in all industry	Differences d_+ (pos)	d_- (neg)
A	55	15	+40	
B	10	15		−5
C	5	5	0	0
D	20	30		−10
E	10	35		−25
Total	100	100	+40	−40

as they are the same—and this total is divided by 100 to give the coefficient, L. Here:

$$L = \frac{\Sigma d_+}{100}$$

$$= \frac{40}{100}$$

$$\therefore L = 0{\cdot}4$$

Values always range from 0 to 1, the former signifying complete identity between the two sets of data (i.e., employment in the industry would be distributed very evenly over the regions, and all the differences would be 0) and the latter indicating extreme concentration of the industry in only one region—i.e., localization.

In the example above there is seen to be a heavy concentration of the brewing industry in region A. This particular coefficient can only be compared with its two extreme values, but comparisons with other industries with different coefficients would lend more meaning to the statistic and enable the student to order a set of industries according to their coefficients of localization.

A rather different example may help to illustrate the use of this technique. Table 25 contains information regarding the numbers in employment and those unemployed in June 1971 in the Economic Planning Regions of Great Britain. These are converted into two sets of percentages as shown in columns C and D and the differences between these two sets are calculated. As can be seen, the positive and negative differences amount to the same total,

Table 25 The localization of unemployment in Great Britain, June 1971

Region	A Numbers in employment	B Numbers unemployed	C % of all in employment i.e. % of A	D % of all unemployed i.e. % of B	differences (d) (D–C)	
					d_+	d_-
South-East	7,352,900	136,721	33·97	19·90		−14·07
East Anglia	606,700	17,965	2·80	2·61		−0·19
South-West	1,325,000	37,853	6·12	5·51		−0·61
W. Midlands	2,206,700	61,060	10·19	8·89		−1·30
E. Midlands	1,351,800	38,430	6·25	5·59		−0·66
Yorkshire and Humberside	1,893,000	70,298	8·74	10·23	+1·49	
North-West	2,718,600	101,531	12·56	14·77	+2·21	
Northern	1,229,000	68,097	5·68	9·91	+4·23	
Wales	961,900	39,741	4·44	5·78	+1·34	
Scotland	2,002,800	115,523	9·25	16·81	+7·56	
Total Great Britain	21,648,400	687,219	100%	100%	+16·83	−16·83

(Source: Department of Employment, Statistics Division)

so it does not matter which difference is used in the calculation of L.

Leading from this table, the calculations for L are as follows:

$$L = \frac{16{\cdot}83}{100}$$

$$\therefore L = 0{\cdot}1683$$

On its own this is a rather raw statistic, but the technique is useful for regional comparisons within a given industry, for comparisons of different industries and for showing the changing distribution of employment over a period of time. It is clear, for instance, that the incidence of unemployment is higher in Wales, Scotland and England north of the Midlands. The relatively good employment situation in the South-East Region is also very evident. Comparison with the coefficient of several years later may illustrate the effect of government assistance or may show the results of increasing concentration of unemployment problems in the same region. (The reader is in fact given the corresponding data for June 1976 as part of the exercises in the following chapter. Direct comparison of the two coefficients may provide the basis for fruitful discussion.)

The coefficient of localization is an extremely simple technique. It is versatile too in that it can be applied to a number of problems or features other than those concerning industrial location. One such use might be the percentage production of a certain crop in a given number of regions against percentage production of all crops in the same regions. But the value of L can vary with the number of regions under consideration and so comparisons from country to country cannot be validly made. Also, L gives no indication of an inverse relationship (i.e., when low values of one feature are associated with high values of another).

The reader should note that the same principle is often used in *the coefficient of geographic association*, g, which is also occasionally referred to as the index of similarity. For this measure the procedure is exactly the same as that for the coefficient of localization, but the result is then subtracted from unity.

$$\text{i.e., } g = 1 - \frac{\Sigma d(+ \text{ or } -)}{100}$$

In this case, obviously, the closer the result is to 0, the greater the difference between the two data sets, and the closer to 1, the greater the similarity.

The location quotient. Like the coefficient of localization, the location quotient enables us to measure the degree of specialization in any one activity. It can be applied, though, to a single industry or activity in only one region or town with reference only to the total for the whole country, without requiring all the percentages for all other regions. The quotient is an indication of the extent to which a specific area has more or less than its share of a particular industry.

The location quotient is given by:

$$LQ = \frac{RI}{RE} \div \frac{NI}{NE}$$

where:

RI = the region's or town's employment in the selected industry;

RE = the region's or town's employment in all industries;

NI = the national employment in the selected industry;

NE = the national employment in all industries.

Thus, if region A of country B employs a total of 5,000 people in its manufacturing industries, of whom 500 are employed in the paper industry, and if the total employment in manufacturing industry in country B is 200,000, of whom 5,000 are employed in the paper industry, then the location quotient for that industry in that region is:

$$LQ = \frac{500}{5{,}000} \div \frac{5{,}000}{200{,}000}$$

$$= 0{\cdot}1 \div 0{\cdot}025$$

$$\therefore LQ = 4{\cdot}0$$

Location quotients above 1·0 show a surplus of the industry or feature, whilst those below 1·0 show a deficiency. So this result indicates that region A of country B employs more than its share in the paper industry, or

Table 26 The calculation of location quotients

Industry	A	B	C	D	E	F	G	H
Food	62	1,486	44,353	0·061	0·367	0·191	0·32	1·92
Beverages	66	31	3,152	0·065	0·008	0·014	4·64	0·57
Tobacco	—	—	1,200	—	—	0·005	—	—
Textiles	—	—	15,528	—	—	0·067	—	—
Footwear	118	298	28,570	0·117	0·074	0·123	0·95	0·60
Wood and cork	578	398	14,289	0·572	0·098	0·061	9·38	1·61
Furniture	4	24	5,639	0·004	0·006	0·024	0·17	0·25
Paper	—	—	9,174	—	—	0·039	—	—
Printing and publishing	56	209	15,803	0·056	0·052	0·068	0·82	0·76
Leather	—	3	2,474	—	0·001	0·011	—	0·09
Rubber	8	11	3,887	0·008	0·003	0·017	0·47	0·18
Chemicals	—	81	7,343	—	0·019	0·032	—	0·59
Petroleum and coal products	—	177	679	—	0·043	0·003	—	14·33
Non-metallic minerals	29	702	8,641	0·029	0·173	0·037	0·78	4·68
Basic metals	—	—	3,878	—	—	0·017	—	—
Metals	8	92	18,355	0·008	0·022	0·079	0·10	0·28
Machinery	64	290	15,895	0·063	0·072	0·068	0·93	1·06
Electrical machinery	—	28	9,997	—	0·007	0·043	—	0·16
Transport equipment	—	189	13,248	—	0·047	0·056	—	0·84
Misc.	17	33	10,539	0·017	0·008	0·045	0·38	0·18
Totals	1,010	4,052	232,424	1·0	1·0	1·0		

Columns:
- A: Numbers employed in Westland, 1971–2
- B: Numbers employed in Northland, 1971–2
- C: Numbers employed in New Zealand, 1971–2
- D: Westland $RI \div RE$
- E: Northland $RI \div RE$
- F: New Zealand $NI \div NE$
- G: Location quotients for Westland
- H: Location quotients for Northland

that there is a definite concentration of employment in the industry in region A.

Table 26 gives values of the location quotients for manufacturing industries in two regions of New Zealand together with columns showing the necessary steps taken in obtaining the quotients. Westland clearly has concentrations of the beverage industry and of the manufacture of wood and cork products. Northland, on the other hand, specializes in the manufacture of petroleum and coal products and the production of non-metallic minerals.

Such measures have obvious implications for regional planning in any given country, but it should be stressed that one location quotient does not tell the whole story. The quotient here is based on numbers employed, but just as valid would be quotients based on the number of factories, the value of production, average wages, and so on.

Table 27 Percentage employment in manufacturing industry in Wales, 1976

Industry	Manufacturing workforce employed %	Rank	Cumulative %
Engineering	25·7	1	25·7
Food, drink and tobacco	14·9	2	40·6
Metal manufacture	10·9	3	51·5
Textiles	9·4	4	60·9
Paper, printing and publishing	7·3	5	68·2
Shipbuilding and marine engineering	7·0	6	75·2
Vehicles	5·3	7	80·5
Clothing and footwear	5·1	8	85·6
Chemicals	4·7	9	90·3
Timber, furniture	3·3	10	93·6
Bricks, pottery, glass, cement	2·9	11	96·5
Other manufactures	2·6	12	99·1
Coal and petroleum products	0·5	13	99·6
Leather	0·4	14	100·0

(Source: Department of Employment, Statistics Division)

The index of diversification. This is very closely related to the location quotient, but concerns the whole industrial mix of a region rather than just one industry. Table 27 gives the percentage employment for each manufacturing industry in Wales. These are ranked (rank 1 to the highest) and the cumulative percentage employment is then plotted against the rank order as is line A in Fig. 32.

The closer the plotted line is to the diagonal, the more diversified is the industry. In fact, the diagonal represents the line of perfect diversification, where each type of industry employs the same percentage of the workforce as all the other types. Line B in Fig. 32 is included as a hypothetical example of a region whose manufacturing base is dominated by only one or two industries.

Fig. 32 The index of diversification of employment in Wales

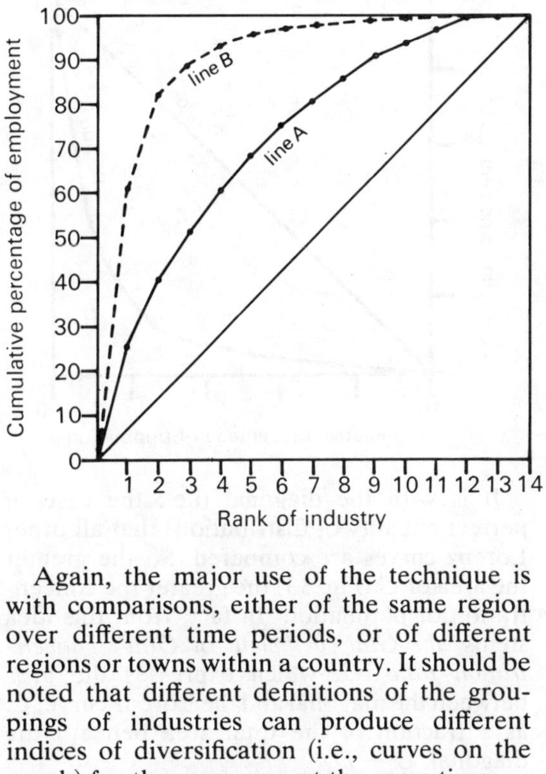

Again, the major use of the technique is with comparisons, either of the same region over different time periods, or of different regions or towns within a country. It should be noted that different definitions of the groupings of industries can produce different indices of diversification (i.e., curves on the graph) for the same area at the same time.

Lorenz curves. These are very similar to the curves drawn for the indices of diversification, and they give a graphic representation of the distribution of any feature such as population, income, industry and others over a given region or country.

To test the evenness of spread of, for instance, a certain population distribution, the cumulative percentage of population is plotted against the cumulative percentage of area, as shown in Fig. 33. Region A in Fig. 33 has a perfectly even distribution of population such that 10% of its area contains 10% of the population, 50% of its area contains 50% of the population, and the same would be true of any other percentage figure given. The Lorenz curve for the population distribution of region A is therefore shown as a diagonal. Region B, on the other hand, illustrates the case in which a large percentage of the population is concentrated in a small percentage of the area.

Fig. 33 Lorenz curve of population distribution

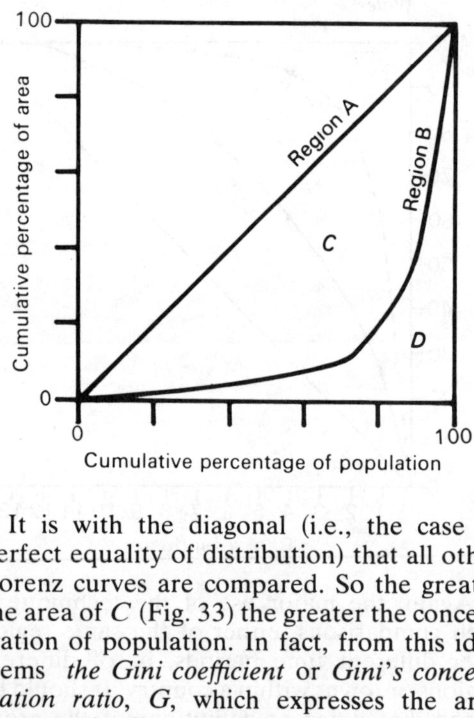

It is with the diagonal (i.e., the case of perfect equality of distribution) that all other Lorenz curves are compared. So the greater the area of *C* (Fig. 33) the greater the concentration of population. In fact, from this idea stems *the Gini coefficient* or *Gini's concentration ratio*, *G*, which expresses the area between the diagonal and the Lorenz curve, *C*, as a fraction of the total area beneath the diagonal, *C* + *D*.

$$\text{i.e., } G = \frac{C}{C + D}$$

A high concentration of population in a small area yields a high Gini coefficient; conversely, the lower the Gini coefficient the more evenly spread the population.

Fig. 34 illustrates Lorenz curves of population distributions for selected countries. The areal units (provinces, states or administrative regions) of each country are arranged in order of decreasing population. The cumulative populations and cumulative areas are then totalled and converted into percentages. Clearly, from Fig. 34, Yugoslavia has a very low Gini coefficient whilst Uruguay's is high. Further examples are included in the exercises in the next chapter. It is also acceptable to arrange the areal units in order of decreasing population densities rather than population numbers.

Fig. 34 Lorenz curves of selected population distributions

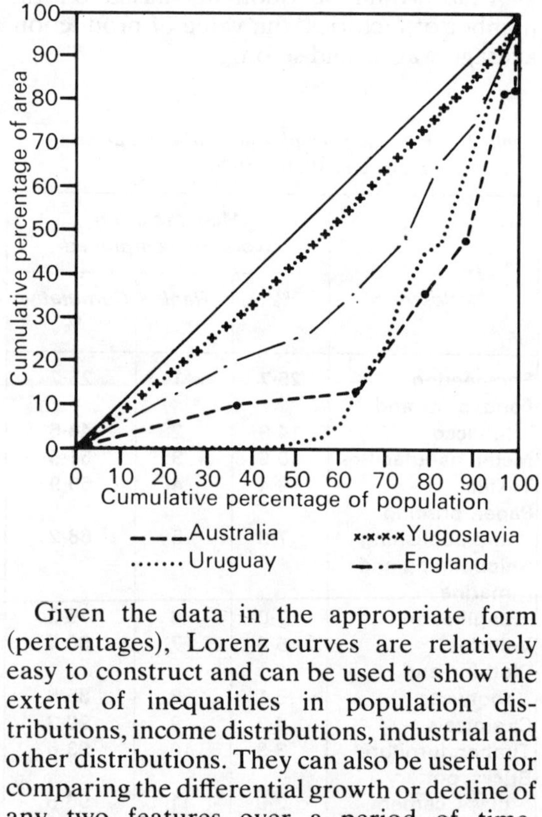

Given the data in the appropriate form (percentages), Lorenz curves are relatively easy to construct and can be used to show the extent of inequalities in population distributions, income distributions, industrial and other distributions. They can also be useful for comparing the differential growth or decline of any two features over a period of time, although it must be remembered that a constant rate of growth over a number of regions will produce the same curve at the end of the time period as at the beginning. Another advantage of this technique is in the clear visual impression of the differences between the two or more countries or regions under study. Nevertheless, the method is a little rough and ready and is not a very powerful descriptive device.

Other measures, ratios and techniques of describing areal distributions exist. Indeed, the shape of regions themselves, a factor that often has much bearing upon the nature of areal distributions, can be measured and described numerically. The aim of this chapter, though, has been to introduce the reader to just a few of the more basic techniques. These can be developed more fully by the teacher if time allows, or if not, the inquisitive student is referred to the list of further reading given at the back of this book.

17 Exercises

Fig. 35 Settlement pattern in part of the Falcon region of Venezuela

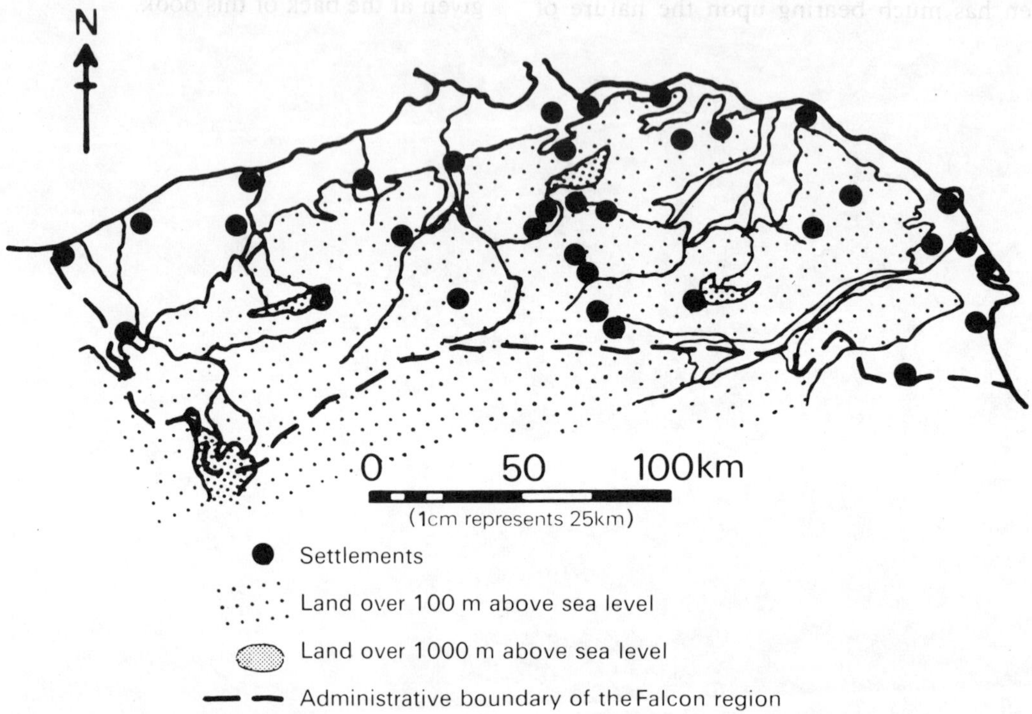

1 Refer to Fig. 35, which shows the distribution of settlements in the major part of the Falcon region of Venezuela. The region is made up largely of a dissected plateau with a narrow coastal plain. It is sparsely populated, semi-arid and covered mostly with scrub vegetation.
(a) Measure the distance (in km) from each settlement to its nearest neighbour.
(b) Given that the area shown on the map covers 25,000 square kilometres, calculate the nearest neighbour statistic for this region.
(c) Using Fig. 30, assess the significance of this result.
(d) With the help of the topographic information shown on the map and climatic information from an appropriate atlas or text book, suggest a few factors that may have influenced this distribution.

2 Refer to Table 28, which gives the numbers employed and the numbers unemployed in June 1976 for the regions of Great Britain along with the regional percentages.
(a) Calculate and tabulate the differences between the percentages shown in columns C and D.
(b) Total either the positive or negative deviations and calculate the coefficient of localization of unemployment.
(c) State what this coefficient, and the calculations involved in it, show about unemployment in June 1976.
(d) Comment upon the differences between this result and that shown in Table 25.

3 Refer to Table 29, which shows employment figures for Greater London, the South East and Great Britain.

Table 28 Numbers employed and unemployed in Great Britain, June 1976

Region	A *Numbers in employment*	B *Numbers unemployed*	C *% of total in employment*	D *% of total unemployed*
South-East	7,246,800	307,923	32·88	24·10
East Anglia	669,500	33,584	3·04	2·63
South-West	1,513,600	97,639	6·87	7·64
West Midlands	2,186,100	126,937	9·92	9·93
East Midlands	1,496,800	74,243	6·79	5·81
Yorkshire and Humberside	1,967,500	115,756	8·93	9·06
North-West	2,637,700	199,119	11·97	15·58
Northern	1,254,700	104,753	5·69	8·20
Wales	995,300	73,796	4·52	5·77
Scotland	2,071,000	144,134	9·39	11·28
Total: Great Britain	22,039,900	1,277,884	100	100

(Source: Department of Employment, Statistics Division)

Table 29 Employment in Greater London, the South-East and Great Britain, June 1976

Industry	*GLC*	*S.E. (inc. GLC)*	*Great Britain*
	(figures in thousands)		
Agriculture, forestry and fishing	1·9	79·2	381·6
Mining and quarrying	3·7	11·8	345·6
----------	----------	----------	----------
Food, drink and tobacco	86·3	158·7	690·6
Coal and petroleum products	2·5	10·7	37·4
Chemicals	49·1	123·6	420·7
Metal manufacture	72·6	148·7	988·5
Engineering	239·6	589·7	1,796·2
Shipbuilding and marine engineering	3·9	37·9	175·4
Vehicles	54·8	196·4	733·0
Textiles	10·2	20·3	479·7
Leather	6·6	10·0	39·7
Clothing and footwear	51·7	77·1	363·6
Bricks, pottery, glass, cement	12·3	48·7	258·0
Timber, furniture	40·3	91·4	259·0
Paper, printing and publishing	122·6	239·2	535·6
Other manufacturing	41·6	98·5	321·3
----------	----------	----------	----------
Construction	178·7	366·3	1,269·2
Gas, electricity, water	51·9	107·6	342·8
Transport and communication	395·8	614·6	1,452·6
Distributive trades	493·0	957·0	2,669·3
Insurance, banking, finance and business	425·7	586·5	1,087·4
Professional and scientific services	591·9	1,266·6	3,559·1
Miscellaneous services	422·6	800·5	2,252·2
Public administration and defence	349·7	604·4	1,580·7
Total manufacturing	794·0	1,851·0	7,098·6
Total all industries and services	3,709·4	7,246·8	22,047·7

(Source: Department of Employment, Statistics Division)

(a) Calculate the location quotients for each industry for both the GLC (with respect to Great Britain) and for the South-East region.
(b) Discuss the results found in (a) above.
(c) What further data and methods of analysis would help to extend a study of the industrial structure of these two areas?

4 Refer again to Table 29.
(a) Calculate the percentage employment in each industrial activity within the GLC.
(b) Rank these percentages. (Rank 1 to the highest.)
(c) Calculate the cumulative percentages for each rank (from 1 to 24).
(d) On a graph, plot the cumulative percentages against the rank to obtain the index of diversification for the GLC.
(e) Draw in the diagonal on your graph and comment upon the difference between this and the index of diversification plotted in (d) above.

5 Refer to Table 30.
(a) By descending order of population, calculate the cumulative percentages of area and population for the regions of both the countries listed in Table 30.
(b) Construct Lorenz curves of population distribution for the two countries, including on your graph the diagonal.
(c) Compare the two Lorenz curves drawn here with those shown in Fig. 34. Briefly try to account for the differences between the curves.

6 Consider how and why the Lorenz curve of England's population distribution (Fig. 34) might change if the areas and populations of Wales and Scotland were included in the data from which this curve was drawn.

Table 30 Population and area of selected countries, by regions

Costa Rica			Brazil		
Region	Area ('000 sq. km)	Population ('000)	Region	Area ('000 sq. km)	Population ('000)
San José	4·9	738·2	South-East	924·9	45,312·8
Alajuela	9·5	345·7	North-East	1,548·7	32,031·5
Puntarenas	11·3	235·0	South	577·7	19,258·2
Cartago	2·6	217·7	Central West	1,879·5	6,328·0
Guanacaste	10·4	191·2	North	3,581·2	4,214·7
Heredia	2·9	141·7			
Limón	9·3	124·2			
Total	50·9	1,993·7	Total	8,512·0	107,145·2

18 Time series analysis

Geographers often use data that appear in time series to explain many features of present distributions. Many geographic theories incorporate the concept of equilibrium—a condition of balance that will ultimately be achieved—although in the past these have generally depended upon the notion of static equilibrium. More recently, however, the idea of temporal changes has been introduced to many geographic theories. A classic example is the theory of dynamic equilibrium in land-form formation. 'Dynamic equilibrium introduces the concept of temporal changes in the formative forces so that the system does not revert to a single point of balance but constantly is changing its position of equilibrium.' (M. T. Daly in *Techniques and Concepts in Geography*, Nelson.)

So, '. . . of late, there has been a small, but growing, realization of the need to incorporate dynamic elements into geographic analyses . . . .' (M. T. Daly in *Techniques and Concepts in Geography*.) The inclusion of a dynamic element has also, of course, become very important in the practice of planning. Indeed, the prediction of a final state, condition or population in 5, 10 or 20 years is often unsatisfactory, and planners attempt to predict the changes that may occur on the factors influencing the present trend.

Interpolation and extrapolation. It could be said that, because of the extremely intricate and complex interactions between all factors influencing trends, it is impossible to forecast, predict or project accurately. There is of course an element of truth in this, but that does not negate the point of planning or prediction. It is very necessary, though, for the planner, geographer, researcher or student to be well aware of the different types of projection and of the assumptions behind them.

A projection can be either interpolated or extrapolated. Interpolation is the prediction of a certain point in time from within the range of data. Fig. 36 illustrates this graphically. The extreme points of the time series are t and $(t + n)$. Given data for these two points, a value for time $(t + a)$ anywhere between the two extremes can be interpolated.

Fig. 36 Interpolation

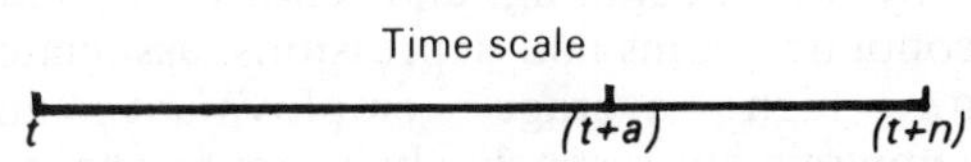

$(t+a)$ = the point predicted

Extrapolation is the prediction of a certain point in time outside the range of data. Fig. 37 illustrates this graphically. Again the extreme points of the time series are t and $(t + n)$. Given data for these two points, a value for time $(t + x)$ can be projected outside this range—i.e., extrapolated.

It should be clear from the very brief outline above that extrapolation is somewhat less reliable than interpolation. Of necessity, however, planning generally has to extrapolate present trends from time series data into the future.

Before outlining some of the methods of describing and analysing time series, some basic elements of time series data should be understood.

Trends and fluctuations. Time series have four components:

(a) *The general trend or secular variation.* This refers to the long-term trend shown in any time series. For instance, in January 1926 there were 9·8 million people in employment in Great Britain. In November 1936, the

Fig. 37 Extrapolation

$(t+x)$ = the point predicted

figure was 11·1 million. Of course this growth showed many variations between those two dates, but the general trend was one of growth.

(b) *Seasonal variations.* The meaning of these is self-evident, but an example may aid understanding. The building industry is generally slack in winter months and therefore employment in this industry is relatively slow during this season. Similarly, the tourist industry experiences a seasonal variation in employment.

(c) *Cyclical variations.* These are similar to seasonal variations, but generally concern a longer time scale, or at least a different time scale that is not rigidly confined by a season or by the beginning and end of a year. Economic booms and depressions, associated with high average employment and unemployment respectively, seem to occur in a cyclical manner. 1929 was a 'boom year', a period of good trade, but was followed by a depression which reached its lowest point of employment in 1932. This trend then reversed, with employment increasing after that year—apart from seasonal variations.

(d) *Special variations.* These are sometimes called residual variations. They are the result of purely random events. They can be irregular or once-and-for-all events that are unpredictable, but which have an influence upon the general trend. In the example used so far, a relevant case might be a general strike or a political crisis.

The easiest method of describing time series data is the simple line graph.

The simple line graph. A series of values is plotted against a time scale on the graph and the resulting points are joined with a continuous line. Fig. 38 shows the new registrations of cars in Great Britain for 15 consecutive months as an example of the simple line graph.

A general trend is not always discernible from a simple graph, because of the immediate and often sharp fluctuations from day to day, month to month or year to year. Rainfall data are notoriously erratic, and comparisons between stations difficult to make, particularly on a daily basis.

Freehand curve fitting. This is rather less attractive than it sounds in practice! It involves the fitting of a general trend line to a series of points shown on a simple graph. It is done purely by eye and as quickly as possible. It is only approximate and is very subjective—a different line or curve would be fitted to the same data sets by different people. Its merit is in the speed with which it is accomplished.

Fig. 38 New registrations of cars in Great Britain

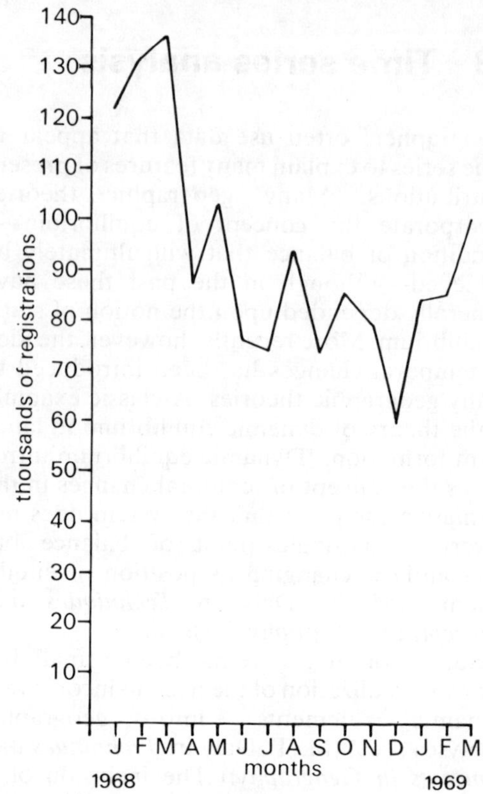

The semi-average method. Again, this method is rather low on accuracy, although a little better than the last, but is relatively simple and quick to use. The data set is divided into two halves, separated by the median. The arithmetic means of both halves are then calculated, and plotted on the graph at the centre of the time scales that they represent. If you join the two points plotted, this should reveal a general trend, as shown by the semi-average trend line in Fig. 39. Variations on this theme can be played by dividing the time scale into thirds, quarters or other fractions—the tri-average trend line is also shown on Fig. 39.

The greater the number of divisions of the original data set, the closer the trend line

Fig. 39 The semi-average and tri-average trends

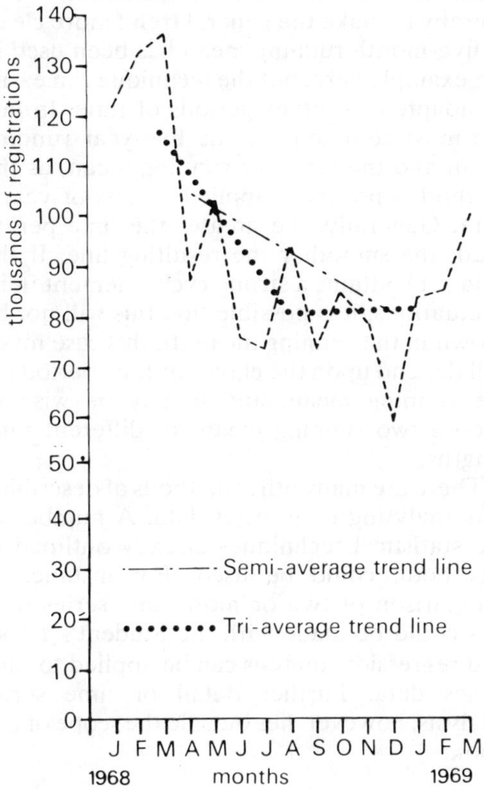

resembles the original line, and therefore the less likely it is to show a general trend.

Running means. As with the previous two techniques, a running mean is used to smooth out great variations so that only the major fluctuations are shown. To this end they are more successful than the previous methods, but their calculation requires rather more time and effort.

Assume that a five-month running mean is being used to smooth out the variations and to show the general trend in the time series of new car registrations in Britain. The first value is the arithmetic mean of the first five items in the data set. This value is plotted against the median month of the first five items. The second value is obtained by calculating the arithmetic mean of the second, third, fourth, fifth and sixth items in the data set (132, 136, 87, 103, 76)—i.e., by dropping the first item and including the sixth in the calculation. This second value is plotted against the fourth item in the data set on the graph, or against the mid-point item of the five items it represents. To calculate the third value, the second item is dropped and the seventh taken into the calculation, and so the process is repeated until the final item in the data set is used. The running means for the example used thus far in this chapter are shown in Table 31. These values are plotted in Fig. 40.

Table 31 Calculation of the five-month running mean

Month	*No. of new registrations (000's)*	*Five-month running mean*
J	122	
F	132	
M	136	116
A	87	107
M	103	95
J	76	87
J	74	84
A	94	80
S	73	81
O	85	78
N	79	76
D	59	78
J	84	81
F	85	
M	100	

Fig. 40 The running mean

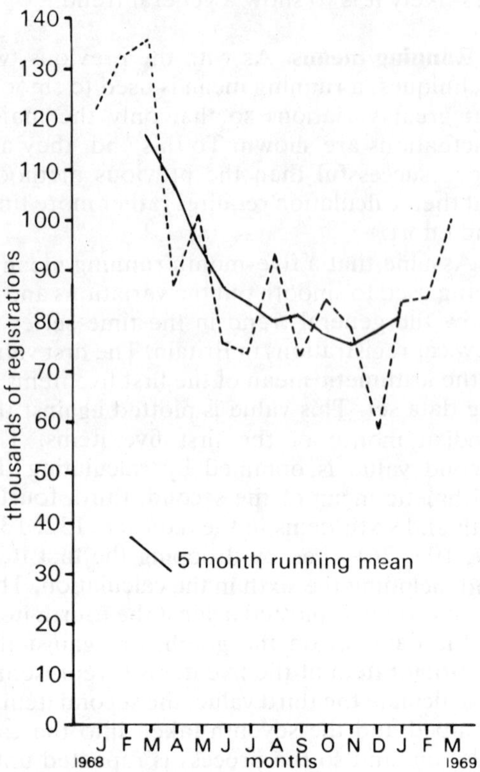

The aim of the running mean is to eliminate seasonal or short-term fluctuations and thereby to make the general trend more clear. A five-month running mean has been used in the example here, but the technique can easily be adapted for other periods of time. In fact the most common are the five-year running mean and the ten-year running mean, as the method is normally applied to sets of yearly data. Generally, the greater the time period used, the smoother the resulting line. If the data set contains a strong cyclic element in its fluctuations, it is possible that this will not be shown in the running mean. In this case much will depend upon the choice of time period for the running mean, and it may be wise to choose two running means of different time lengths.

There are many other methods of describing and analysing time series data. A number of the statistical techniques already outlined in this book could be used. For instance, a comparison of two or more time series data sets could be made with the Student's t test, and regression analysis can be applied to time series data. Further detail of time series analysis, however, lies outside the scope of this book.

19 Exercises

1 (a) Re-plot the data shown in Figure 38.
(b) Fit a freehand curve to the resulting line graph.
(c) Calculate the semi-average trend for the same data and plot this on the graph used for (a) and (b) above.
(d) Comment upon the differences between the freehand curve and the semi-average trend line shown on the graph.

2 Table 32 shows the average pollution levels in Sheffield and the number of complaints received by the Public Health Department.
(a) Plot both data sets shown in the table as a simple line on the same graph.
(b) Apply a five-year running mean to both data sets and mark the resulting trend lines on the graph.
(c) In your view, do the five-year running means help the reader to interpret the data?

Table 32 Pollution levels and number of complaints—Sheffield

Year	Pollution levels ($mg/m^2/day$ of solids)	Number of complaints regarding pollution received by Public Health Department
1950	218	112
1951	215	93
1952	224	90
1953	225	160
1954	248	143
1955	241	159
1956	243	158
1957	190	178
1958	188	175
1959	119	196
1960	164	188
1961	150	199
1962	151	201
1963	147	192
1964	147	252
1965	138	157
1966	137	271
1967	145	247
1968	128	252
1969	147	289

20 Network analysis—physical geography

Analysis of those properties of networks that can be quantified has recently been taken up as an alternative to qualitative descriptions of drainage networks in physical geography and transport networks in human and regional geography.

In physical geography the major use of the technique of network analysis is in the map study of drainage basins and channels—although it is worth remembering that the most extensive and effective map analysis is accompanied by field survey. The terms normally used to describe drainage patterns, such as 'dendritic', 'trellised', 'angular', 'radial', etc., convey a meaning but are imprecise and open to widely varying interpretations. It is in order to avoid or minimize these faults that statistical analysis of drainage networks is most useful. A more formal statement of the aims of the technique is given below.

Aims. The aims are to:

1 improve understanding of the organization of drainage basins;

2 facilitate classification of drainage basins;

3 tie the channel and basin together;

4 derive meaningful generalizations, and possibly laws, about the relationship between channel elements and basin elements, by the use of a numerical and objective approach;

5 facilitate comparisons of different drainage basins.

Stream ordering. The technique of drainage network analysis is based on the ordering of streams or channels in an hierarchical manner. First order streams are small tributaries and are the lowest in the hierarchy. The greater the order of the stream, the higher it stands in the hierarchy, and therefore the larger the stream. As stream order increases the size and volume of the stream also increases.

Different methods of stream ordering exist, but the most commonly used is that devised by Strahler.* (Other methods are outlined in *Networks in Geography*, by Dalton, R. *et al.*, Philip.) This is best shown in diagrammatic form, as in Fig. 41, on which the stream orders are marked.

Fig. 41 Stream ordering after Strahler

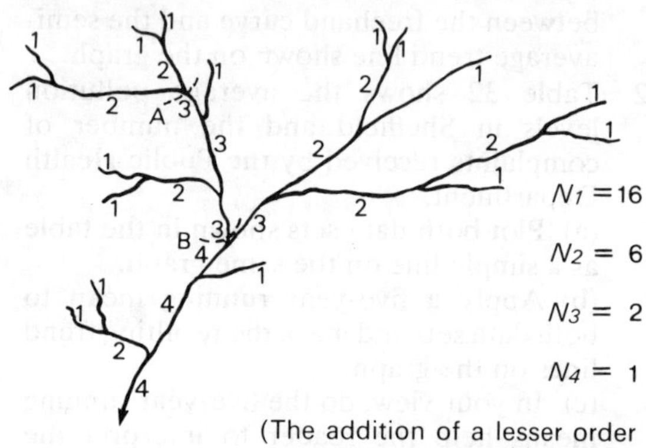

(The addition of a lesser order stream to, say, a third order stream does not increase the number of third order streams. There is only one third order stream between points A and B on this map.)

The smallest finger-tip tributaries are designated as first order streams. Where two first order streams join, the combined stream is second order. A third order stream commences only where two second order streams meet. If a first order stream joins a second or higher order stream the order of the combined stream is not changed. The order is only increased where two streams of the same order meet; so where a third order stream joins a fourth order stream, the combined stream remains of the fourth order.

Stream ordering is only the start of drainage basin analysis. The whole procedure is set out below.

Procedure. 1 Third or fourth order basins are generally the most appropriate for study. Where possible, it is sensible to select

* Strahler, A. N.: *The Earth Sciences* (Harper and Row, 1965).

drainage basins for which climatic data (particularly rainfall) and structure are known.

2 Define the drainage network according to the blue lines on the map. For later stages in the analysis it is wise to trace the network.

3 Assign an order to each stream in the network (according to Strahler).

4 The following information should be collected and entered onto a data sheet such as that shown in Table 33:

(a) the number of streams of each order;

(b) the total length of streams of each order, in kilometres (ideally, measured with a French opisometer, but a thin string could be used);

(c) the total fall of streams of each order (to the nearest five metres) by interpolation between the contour lines;

(d) the area of the drainage basin. (It is recommended that the watershed of the whole basin should be drawn on the tracing paper overlay at an early stage in the analysis. Fig. 42 shows that the watershed should be drawn orthogonal to the contours, and should be drawn in for the lower order basins as well as the major drainage basin. If spot heights occur on a wide watershed, these should be taken as

Table 33 Data record sheet—drainage basin analysis

Name of drainage basin: ....

Grid reference at mouth of basin: ....

O.S. sheet number: .... Scale: ....

Mean annual rainfall: ....

Major type of lithology: ....

Notes on vegetative cover and soil: ....

....

Number of streams/channels of each order (N)	N_1	Bifurcation ratio	$\frac{N_1}{N_2}$ ....
	N_2		$\frac{N_2}{N_3}$ ....
	N_3		$\frac{N_3}{N_4}$ ....
	N_4		
	$\Sigma N =$	Average bifurcation ratio	
Length of streams/channels of each order (L)	L_1 ....km	Average length	$\bar{L}_1$ ....km
	L_2 ....km	$\bar{L} = \frac{\Sigma L_0}{\Sigma N_0}$	$\bar{L}_2$ ....km
	L_3 ....km		$\bar{L}_3$ ....km
	L_4 ....km		$\bar{L}_4$ ....km
	$\Sigma L =$ ......km		$\bar{L}$ ....km
Fall of streams/channels of each order (F)	F_1 .... m	Gradient	$\bar{G}_1$ ......m/km
	F_2 .... m	$\bar{G} = \frac{\Sigma F_0}{\Sigma L_0}$	$\bar{G}_2$ ......m/km
	F_3 .... m		$\bar{G}_3$ ......m/km
	F_4 .... m		$\bar{G}_4$ ......m/km
	$\Sigma F =$ ...... m		$\bar{G}$ ......m/km

Area of basin $A =$ ......sq. km

Length of perimeter $P =$ ......km

Drainage density $Dd = \frac{\text{Total length of streams}}{\text{Area of basin}} = \frac{\Sigma L}{A}$

Drainage frequency $Df = \frac{\text{Total number of streams}}{\text{Area of basin}} = \frac{\Sigma N}{A}$

Relative relief $Rr =$ Height range . length of longest axis

Fig. 42 First, second and third order watersheds

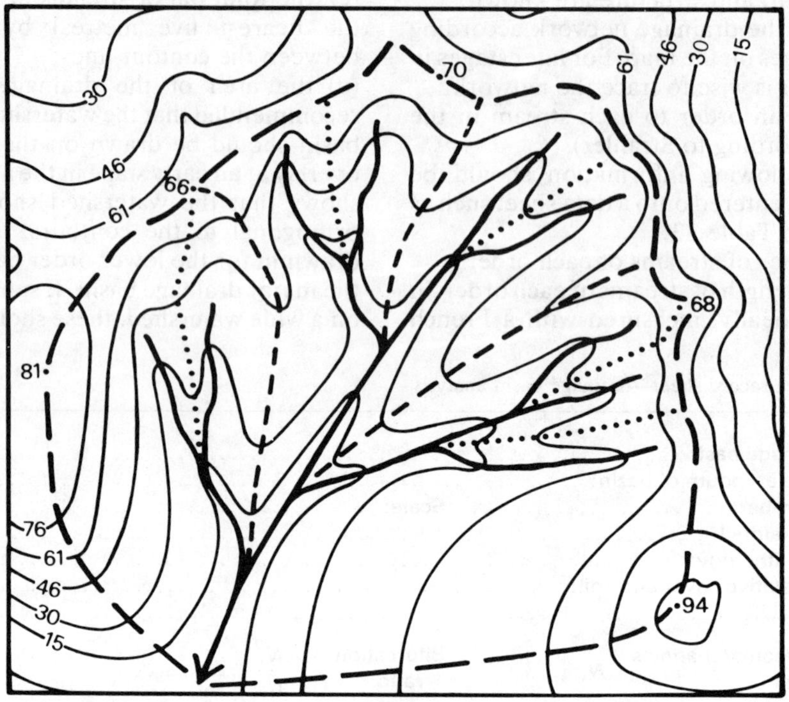

1 : 25,000

Watershed of the third order basin
Watershed of the second order basins
Watershed of the first order basins
15 Contours – height in metres

(In order to delimit the extent of all the first order basins it is necessary to add more first order watersheds to the map.)

representing summits. Squared tracing paper can be used to measure the area of the drainage basin);

(e) the length of the perimeter of the drainage basin (in kilometres).

5 When collated, these simple measurements yield a considerable amount of information about the drainage basin, and lead on to measures such as the bifurcation ratio, the gradient, the drainage density and others which are listed in Table 33. A short explanation of some of these measures may be required:

(a) The bifurcation ratio is the ratio of the number of streams of a given order to the number of streams of the next highest order. One would expect this to decrease as the stream order increases, and its usefulness is found in the comparison of different drainage basins, especially in areas of differing rock types.

(b) The average gradient, in metres per kilometre, is calculated from the total fall and total length of streams of a given order. This would be a helpful measure to use with studies of the long profiles of rivers.

(c) The drainage density is given by the total length of streams of a given order basin divided by the total area of that same basin. As with the following two measures, the use of this is found mainly in comparisons of different drainage basins. Values can be expected to

vary with differing rock types and different climatic conditions.
(d) The drainage frequency is given by the number of streams in a given order basin divided by the total area of that same basin.
(e) The relative relief is the range in altitude of the basin multiplied by the length of the longest axis of the basin. The greater the range in altitude, the greater will be the relative relief.

Most of these measures would of course vary with structure, rock type, rainfall and other factors. Comparisons of drainage basins should attempt to incorporate as many of these factors as possible into the study.

Conclusions from the measurements. The major advantage of this type of analysis is that our conclusions can be precise and leave less room for biased interpretations.

The most obvious conclusions concern the relationships between the stream characteristics and the basin characteristics. So, at the simplest level, we ask whether gradient increases or decreases as stream order increases, whether stream length increases or decreases as stream order increases, and so on. Large-scale map survey (without fieldwork) can establish these properties and relationships.

It may be relevant to measure channel elements rather than stream elements, particularly in areas of limestone or chalk. Without the guide of the blue lines (owing to the dry valleys) the initial delimitation of the channels is, of course, more difficult. Other characteristics, such as valley width, may be important in glaciated areas, and these too could be incorporated into the analysis.

Precise expressions of the properties and relationships within a drainage basin help in comparisons of different drainage basins. Comparative studies are particularly worthwhile where data concerning external factors (e.g., rock type, rainfall, vegetative cover, etc.) are available.

The conclusions drawn from this type of analysis could be more thorough and searching if they were supported by fieldwork. This would enable us to include into the analysis characteristics such as stream width, stream depth, speed of flow and the load.

Fig. 43 A drainage network shown on two different scales

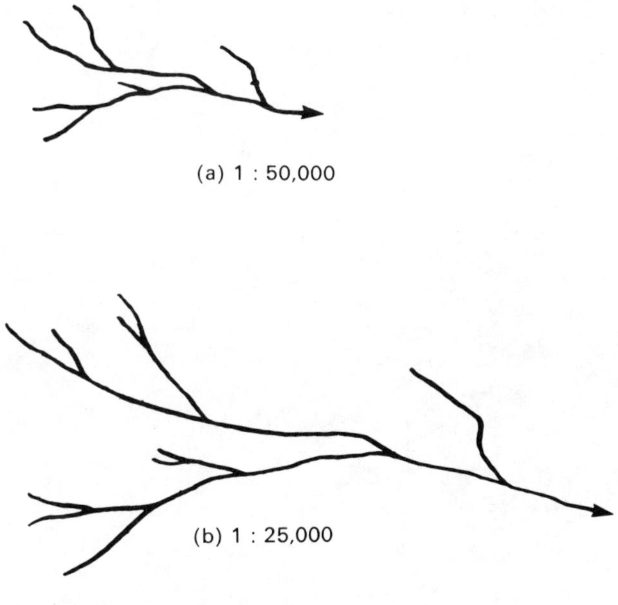

(The reader is left to discover the differences in stream ordering for him/herself.)

Limitations. (1) The depth of analysis is restricted by the amount of information given on the map. Map analysis does not replace fieldwork.

(2) Other basic data (the external factors) are necessary if the conclusions are to be really worthwhile. Third- and fourth-order basins are the most suitable for this type of map analysis, but it is unfortunate that there are few of these for which this extra information is available.

(3) Differences in map scale always present a problem. It would be misleading to compare a drainage network shown on a scale of 1:25,000 with one on a scale of 1:50,000. Fig. 43 depicts the same network of streams at two different scales and clearly illustrates the reason for this non-comparability between map scales.

(4) The measurement of stream lengths on a map is not an easy task, and inevitably human error will creep in here. Human error, of course, creeps into most measurements, but the danger with this exercise is one of excessive error.

(5) Many rivers and streams experience conditions that vary with the season of the year. Periods of heavy rainfall can quickly produce new first-order tributaries that have not been surveyed and are not shown on the map. Naturally, these alter the order assigned to the more permanent channels downstream.

Awareness of these limitations is essential when this technique is employed; but they do not negate the use nor the validity of the technique. If this type of analysis can tighten our previously loose categorization of drainage basins and can improve and support our qualitative descriptions of stream, channel and network characteristics, then it is both worthwhile and necessary.

21 Exercises

The first four exercises in this chapter refer to the drainage basin shown in Fig. 44.

1 Assign an order to each stream in the drainage basin. (It may be advisable to make your own tracing of the network.)

2 Construct your own data sheet as in Table 33. Make all the necessary measurements of the network of streams and enter the measurements onto your data sheet. (This is probably best done as a group task, dividing the measurements to be made amongst a number of students.)

3 In one sentence each, state the relationships found between the following elements:
(a) stream order and the number of streams;
(b) stream order and stream length;
(c) stream order and the bifurcation ratio;
(d) stream order and the fall in altitude;
(e) stream order and the average gradient.

4 Calculate:
(a) the drainage density;
(b) the drainage frequency;
(c) the relative relief.

5 (a) Select a drainage basin of the same order as the Cwm Orky and for which you have a map on the same scale—i.e., 1:25,000.
(b) For the drainage basin selected, repeat the measurements, calculations and tabulation that you made for the Cwm Orky.
(c) Compare your results with those obtained for the Cwm Orky. Do the relationships found for the selected drainage basin support those already found for the Cwm Orky? Can any differences in the results be explained by knowledge of the rock type, structure, climatic factors, vegetative cover and/or man's influence?

6 Outline a fieldwork programme designed to:
(a) confirm or disprove the findings of your map analysis for a selected drainage basin;
(b) expand on the relationships and results already discovered through map analysis.

7 Can you suggest any other features for measurement that could be incorporated into the analysis of:
(a) a glaciated drainage basin?
(b) a drainage basin in a limestone area?

Figure 44 Drainage basin of the Cwm Orky (a Rhondda tributary)

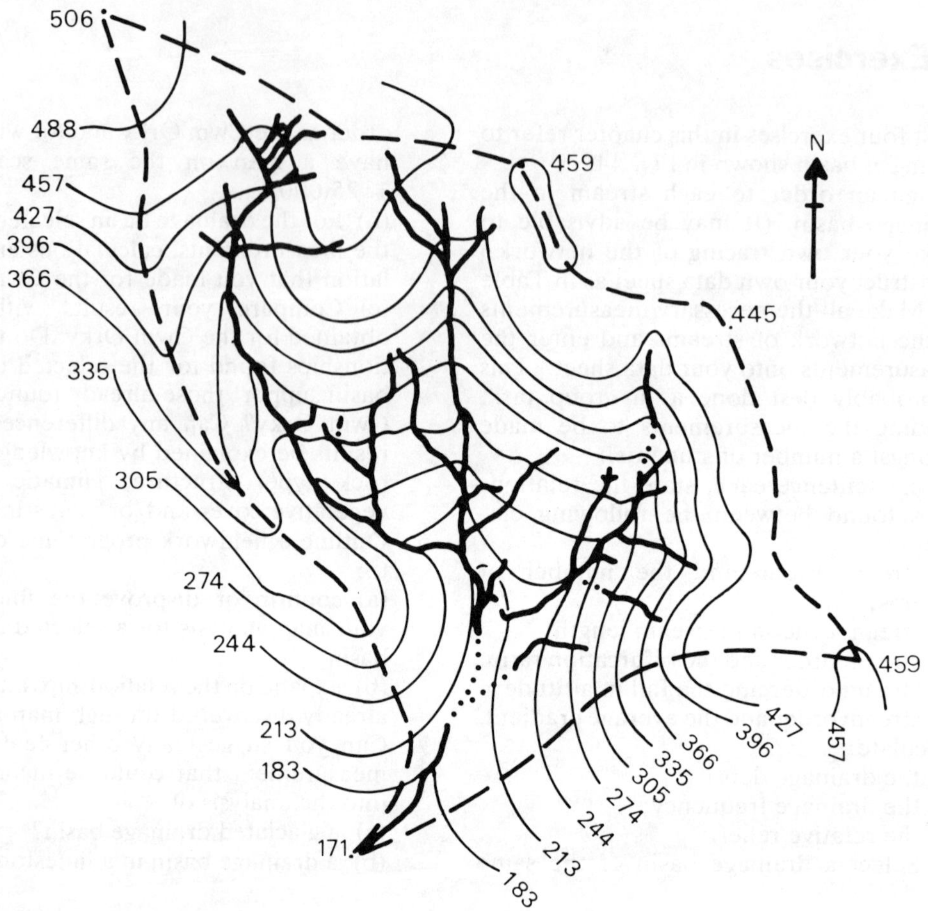

watershed

discontinuity in the stream, although channel continued on the map

contours at 30m intervals

Scale 1:25,000 (4cm represent 1km.)

O.S. map sheet SS 99
Grid reference at mouth of basin 953972
(Consultation of sheet SS 99 would be of help in the analysis.)

22 Network analysis—human geography

The major use of network analysis in human and regional geography concerns transport networks, although business communication networks could also be subjected to this type of analysis.

The technique is based upon the transformation of a transport network to a graphical form, as shown in Fig. 45. This simplifies the network into nodes (v), edges (e) and subgraphs (p)—see below for definitions—and disregards other properties such as distance, node size, orientation, traffic flow, etc. Also omitted at this stage are factors that influence the development of the network such as region shape and size, relief of the land and population distribution.

Fig. 45 Transformation of a network into a graph

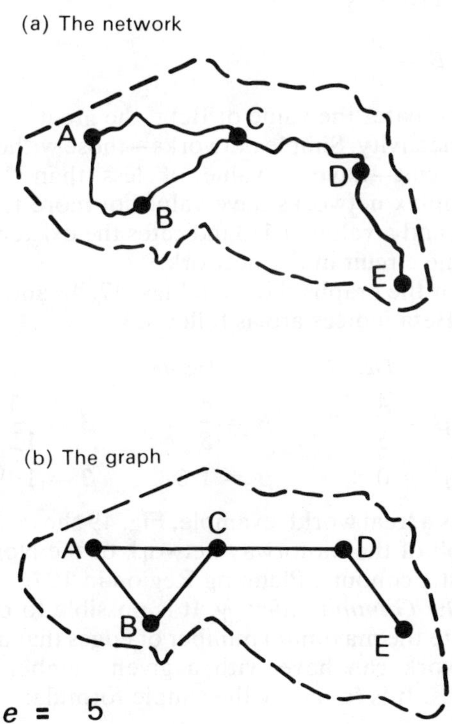

$e = 5$

$v = 5$

$p = 1$

The analysis then uses graph theory (a branch of topology) to derive measures which describe the network or graph. This method of analysis was devised by K. J. Kansky in 1963.* It enables us to:

1 express exact relationships between network properties;

2 establish an order of networks and of particular properties of those networks;

3 use graphs as simple models of the real world to give a visual impression of spatial relationships.

Procedure. The first step in the analysis is to reduce the network to a graph made up of nodes, edges and subgraphs.

The *edges* are the routes or lines joining places. They are often referred to as links, branches or routes, but will here be known as edges and given the notation *e*.

The *nodes* or *vertices* are the places joined by the edges. They are given the notation v and should occur wherever edges intersect and at the ends of edges.

The *subgraphs* are the number of graphs or networks under consideration as unconnected parts of the whole graph or network. They are generally denoted by *p* or *G*. The former is used here.

Fig. 45(a) shows a simple network of roads connecting five places. Fig. 45(b) shows the same network simplified into a graph. The values of the elements of the graph (e, v and p) are given with the diagram.

An example from the real world may help to illustrate this process of breaking down or transforming a network into a graph. Fig. 46(a) shows Guyana's railway network. Fig. 46(b) gives the graphical form of this. A node has been included wherever a town of more than 10,000 population is found on the network or at the end of an edge in the network. These first assumptions about town size are very significant. If a population of 20,000 had been stipulated there would be

* K. J. Kansky, *Structure of Transportation Networks* (University of Chicago, 1963).

Fig. 46 The railway network of Guyana

(a) The network

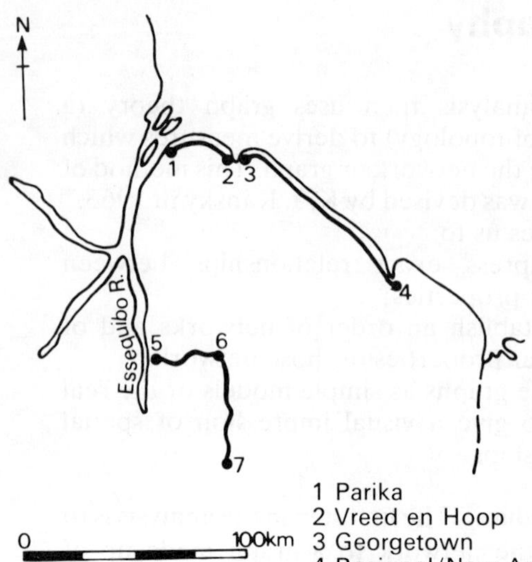

(b) The graph

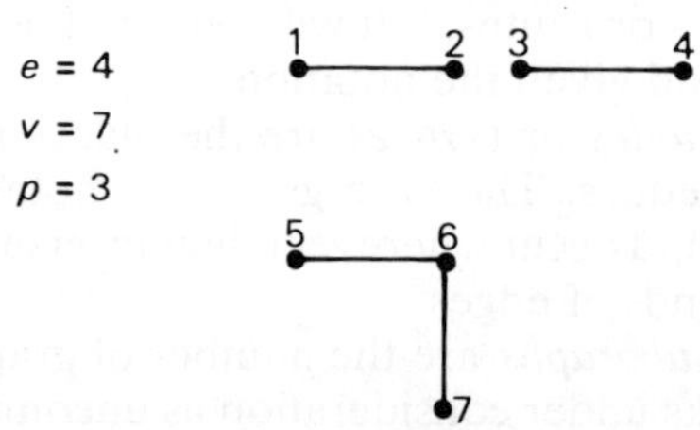

fewer nodes; if 1,000 there would be more. This obviously affects the value of v and could have a strong influence upon further analysis and its results. It is vital, therefore, that these assumptions and definitions of town/node size should be carefully made and clearly stated at an early stage in the analysis.

Measures showing connectivity and accessibility are derived from the basic data for each graph. The term connectivity relates to the whole graph. A highly connected graph is one which has many edges joining the nodes. A poorly connected graph has few edges between the nodes. Figs. 47 and 48 illustrate the notion of low and high connectivity.

Accessibility, on the other hand, refers to a point or points on the graph rather than describing the graph as a whole. Indices of accessibility show how easily a node is reached from all other nodes on the graph. Node E is

Fig. 47 Low connectivity

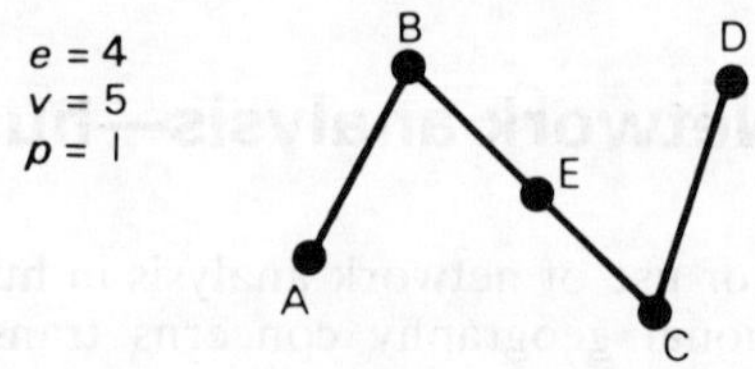

Fig. 48 High connectivity

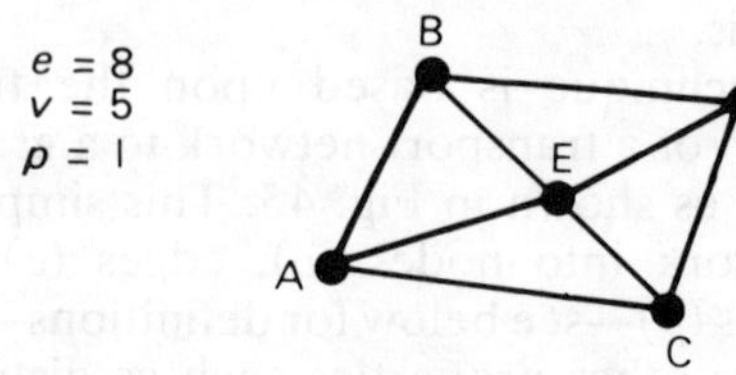

clearly the most accessible node in both Figs. 47 and 48, but node E in Fig. 48 is more accessible than node E in Fig. 47.

Measures of connectivity. The simplest measure of connectivity relates the number of edges to the number of nodes. *The Beta index*, β, is given by:

$$\beta = \frac{e}{v}$$

The greater the value of Beta, the greater the connectivity. Simple networks—those without a circuit—have a value of less than 1·0. Complex networks have values of more than 1·0, and a value of 1·0 indicates the existence of one circuit in the network.

For the graphs shown in Figs. 47, 48 and 49 the Beta indices are as follows:

Fig. 47	*Fig. 48*	*Fig. 49*
$\beta = \frac{4}{5}$	$\beta = \frac{8}{5}$	$\beta = \frac{13}{12}$
$\beta = 0{\cdot}8$	$\beta = 1{\cdot}6$	$\beta = 1{\cdot}08$

As a 'real world' example, Fig. 49 shows the graph of the motorway network of the North West Economic Planning Region in 1976.

The Gamma index, γ. It is possible to calculate the maximum number of edges that any network can have with a given number of nodes. It is found by the simple formula:

$$e_{max} = 3(v - 2)$$

where e_{max} represents the maximum number of edges. This assumes that the maximum

Fig. 49 Graph of the motorway network of the North West Economic Planning Region, 1976

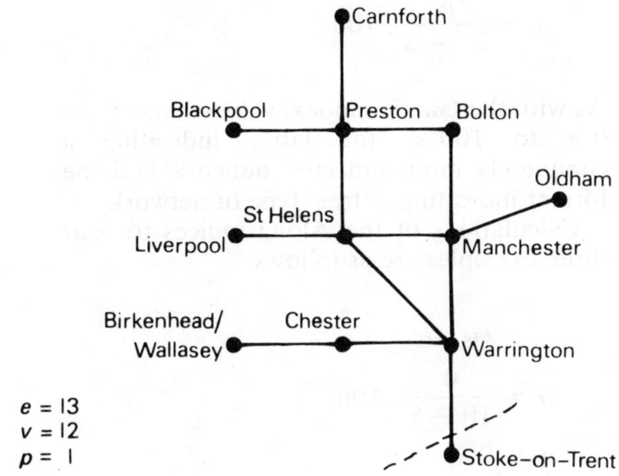

number of edges would be found if every node was linked to every other node on the network.

For the three graphs already shown:

Fig. 47

$$e_{max} = 3(5 - 2)$$
$$e_{max} = 3 \,.\, 3$$
$$e_{max} = 9$$

Fig. 48

$$e_{max} = 3(5 - 2)$$
$$e_{max} = 3 \,.\, 3$$
$$e_{max} = 9$$

Fig. 49

$$e_{max} = 3(12 - 2)$$
$$e_{max} = 3 \,.\, 10$$
$$e_{max} = 30$$

This enables us to calculate a ratio between the actual number of edges and the maximum possible number of edges. This yields the Gamma index and is obviously a measure of connectivity. It is given by:

$$\gamma = \frac{e}{3(v - 2)} \,.\, 100$$

Values range from 0% to 100%, the latter indicating a network that is completely connected. In the three examples used here:

Fig. 47

$$\gamma = \frac{4}{3(5 - 2)} \,.\, 100$$
$$\gamma = \frac{4}{9} \,.\, 100$$
$$\gamma = 44\%$$

Fig. 48

$$\gamma = \frac{8}{3(5 - 2)} \,.\, 100$$
$$\gamma = \frac{8}{9} \,.\, 100$$
$$\gamma = 89\%$$

Fig. 49

$$\gamma = \frac{13}{3(12 - 2)} \,.\, 100$$
$$\gamma = \frac{13}{30} \,.\, 100$$
$$\gamma = 43\%$$

Although this index is independent of the number of nodes in the network and can therefore be used to compare networks of different sizes, it is clear that some factors are not accounted for in the above examples. It seems strange, for instance, that the motorway network of the North West Economic Planning Region appears to have a lower degree of connectivity than the graph shown in Fig. 47.

It is probably wise then to employ more than one index of connectivity when studying networks. The following two indices, for example, incorporate more than just the numbers of edges and nodes into the analysis; they also use the idea of circuits in a network.

The cyclomatic number gives the number of circuits in the network or graph. It is generally denoted by μ, which is open to confusion with the μ that is sometimes used to represent the arithmetic mean. It is given by:

$$\mu = e - v + p$$

For the graphs shown in Figs. 47, 48 and 49, the calculations for the cyclomatic number are as follows:

Fig. 47

$$\mu = 4 - 5 + 1$$
$$\mu = 0$$

Fig. 48

$$\mu = 8 - 5 + 1$$
$$\mu = 4$$

Fig. 49

$$\mu = 13 - 12 + 1$$
$$\mu = 2$$

If an edge was added between nodes A and E in Fig. 47 a circuit would be created joining nodes A, B and E. The cyclomatic number would then be 1.

A high value of the cyclomatic number indicates a highly connected network. It has been shown that there exists a significant relationship between a region's level of economic development and the value of the cyclomatic number of its major transport networks. A little caution is required in the interpretation of this relationship, however, as networks with different types of structures can have the same value of μ.

The Alpha index, α, is the ratio of the actual number to the maximum possible number of circuits in the network. It relates the actual cyclomatic number to the maximum possible cyclomatic number. The latter is dependent on the maximum possible number of edges in the network, which has already been calculated for the Gamma index and is given by:

$$e_{max} = 3(v - 2)$$

Substituting this into the formula for the cyclomatic number, and assuming one connected graph:

$$\mu_{max} = e_{max} - v + p$$
$$\mu_{max} = 3(v - 2) - v + 1$$
$$\mu_{max} = 3v - 6 - v + 1$$
$$\mu_{max} = 2v - 5$$

The formula for the Alpha index is therefore:

$$\alpha = \frac{\mu}{2v - 5} \,.\, 100$$

As with the Gamma index, values range from 0% to 100%, the latter indicating a completely interconnected network and the former indicating a 'tree' type of network.

Calculations of the Alpha indices for our three examples are as follows:

Fig. 47

$$\alpha = \frac{0}{10 - 5} \,.\, 100$$
$$\alpha = 0\%$$

Fig. 48

$$\alpha = \frac{4}{10 - 5} \,.\, 100$$
$$\alpha = 80\%$$

Fig. 49

$$\alpha = \frac{2}{24 - 5} \,.\, 100$$
$$\alpha = 10{\cdot}5\%$$

The student should compare these results with others showing the degree of connectivity—particularly with the Gamma index, as both the Alpha and Gamma indices are expressed as percentages. It is also a worthwhile exercise to add an edge between nodes A and E in Fig. 47 and to calculate the indices of connectivity for the newly formed network.

Accessibility. The measures outlined so far are used to describe the degree of connectivity of a network as a whole. It is possible though to describe the accessibility of any particular node on a network. This is done through the compilation of a shortest path matrix which lists the number of edges used from each node in the shortest paths to all other nodes on the network.

Figs. 50 and 51 give further examples of graphs along with the shortest path matrices for each.

Fig. 50 The South West motorway network, 1976

(a) Graph

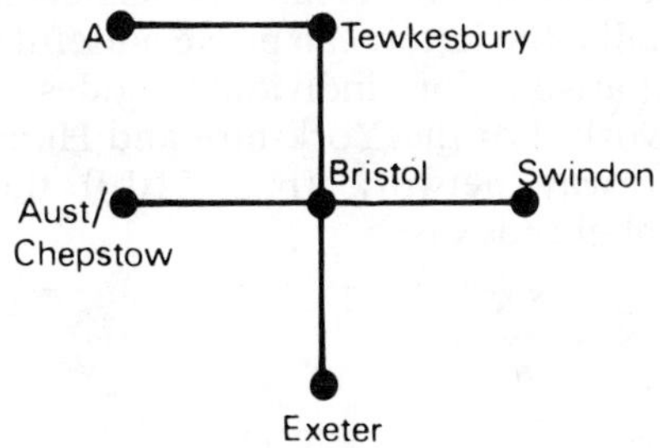

(b) Shortest path matrix

		A	B	C	D	E	F	Total
A		0	1	2	3	3	3	12
B	Tewkesbury	1	0	1	2	2	2	8
C	Bristol	2	1	0	1	1	1	6
D	Aust/Chepstow	3	2	1	0	2	2	10
E	Exeter	3	2	1	2	0	2	10
F	Swindon	3	2	1	2	2	0	10

The Associated Number of any node is the number of edges from that node to the most distant node from it. This measure of distance, however, is in edges rather than kilometres; that is, topological distance. So in Fig. 50 the topologically most distant node from Exeter is node A. The two are separated by three edges in the shortest possible route between them. The Associated Number of Exeter is therefore 3.

From the shortest path matrix the Associated Number for each node is the highest in each row. Fig. 50(b) shows that node A, Aust/Chepstow, Exeter and Swindon all have Associated Numbers of 3, whilst Tewkesbury and Bristol have Associated Numbers of 2. Tewkesbury and Bristol are therefore the most accessible nodes on this network. Leeds/Bradford, Sheffield and Castleford are the most accessible nodes in the Yorkshire and Humberside motorway network (Fig. 51) with Associated Numbers of 3.

Having counted the Associated Numbers for each node, it may be desirable and worthwhile to calculate the mean Associated

Fig. 51 The Yorkshire and Humberside motorway network, 1976

(a) Graph

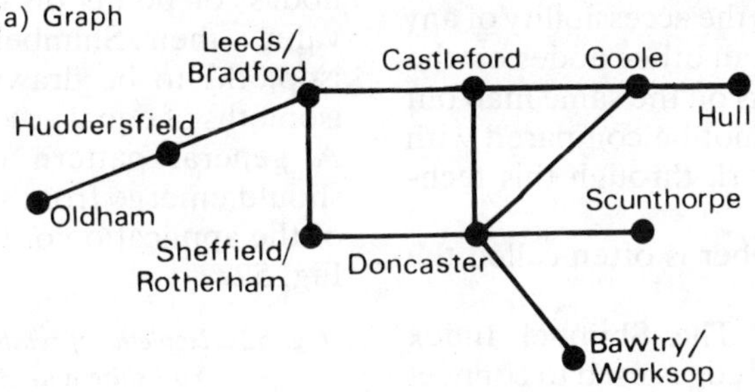

(b) Shortest path matrix

		A	B	C	D	E	F	G	H	I	J	Total
A	Oldham	0	1	2	3	4	5	3	4	5	5	32
B	Huddersfield	1	0	1	2	3	4	2	3	4	4	24
C	Leeds/Bradford	2	1	0	1	2	3	1	2	3	3	18
D	Sheffield/Rotherham	3	2	1	0	1	2	2	2	3	2	18
E	Doncaster	4	3	2	1	0	1	1	1	2	1	16
F	Bawtry/Worksop	5	4	3	2	1	0	2	2	3	2	24
G	Castleford	3	2	1	2	1	2	0	1	2	2	16
H	Goole	4	3	2	2	1	2	1	0	1	2	18
I	Hull	5	4	3	3	2	3	2	1	0	3	26
J	Scunthorpe	5	4	3	2	1	2	2	2	3	0	24

Number for the network. This is calculated in the same way as the arithmetic mean, by totalling all the individual Associated Numbers and dividing by the number of nodes. For the network in Fig. 51:

$$\bar{A} = \frac{\Sigma A}{n}$$

where:

A = the Associated Number
$\bar{A}$ = the mean Associated Number
n = the number of nodes on the network

$$\therefore \bar{A} = \frac{5+4+3+3+4+5+3+4+5+5}{10}$$

$$\bar{A} = \frac{41}{10}$$

$$\bar{A} = 4{\cdot}1$$

This is a useful statistic against which other Associated Numbers can be compared and categorized as either more or less accessible. However, it cannot be used for the comparison of different networks unless those networks have the same number of nodes. Moreover, the accessibility shown by the Associated Number should correctly be referred to as 'relative accessibility' as the accessibility of any node is relative only to all other nodes on the network. Areas or points on the same map but not on the network cannot be compared with the nodes on the network through this technique.

The Associated Number is often called the König number.

The Shimbel Index. The Shimbel Index measures the number of edges used to connect any node with all other nodes on the network by the shortest possible routes. Again, the shortest path matrix contains the necessary information. The summing of all the shortest paths from one node to all others gives the Shimbel Index—the lower the value of the Index, the greater the relative accessibility.

This usually gives a more revealing picture of accessibility than the Associated Number. For example, the Associated Numbers of Tewkesbury and Bristol (Fig. 50) are the same, whereas the Shimbel Index shows Bristol to be more accessible. Whilst there are only three values of the Associated Number (3, 4 and 5) in the Yorkshire and Humberside motorway network (Fig. 51), there are five values of the Shimbel Index (16, 18, 24, 26 and 32.)

The mean Shimbel Index is calculated in the same way as the mean Associated Number and, like the latter, can prove a useful point of comparison for individual nodes on the network. For the Yorkshire and Humberside motorway network (Fig. 51(b)) the mean Shimbel Index is:

$$\bar{S} = \frac{\Sigma S}{n}$$

where:

S = the Shimbel Index
$\bar{S}$ = the mean Shimbel Index
n = the number of nodes.

$$\bar{S} = \frac{32 + 24 + 18 + 18 + 16 + 24 + 16 + 18 + 26 + 24}{10}$$

$$\bar{S} = \frac{216}{10}$$

$$\bar{S} = 21{\cdot}6$$

The values of the Shimbel Indices for nodes on a network can be used as the data base for the construction of an isopleth map. The nodes, or points on the map, are all given a value—their Shimbel Index; values of the isopleths to be drawn are selected; and the isopleths of equal accessibility are then drawn. A general pattern of relative accessibility should emerge from such a map. An example of the application of this technique is given in Fig. 52.

Fig. 52 Isopleths of relative accessibility in the Yorkshire and Humberside motorway network

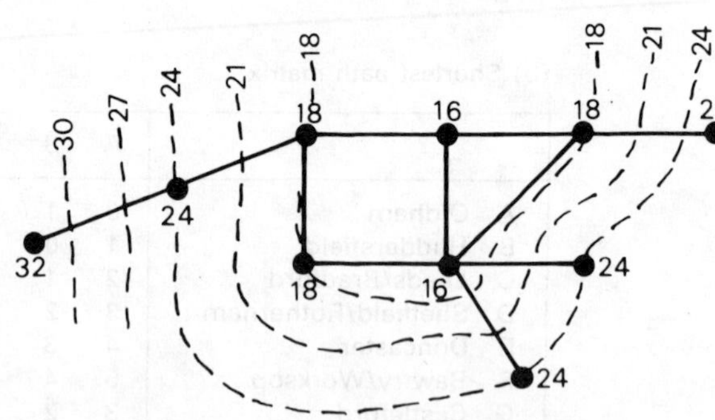

(The Shimbel Index of each node is printed by its side.)
(Refer to Fig. 51)

Fig. 53 Changes in the Eta Index

(a) $e=5$

$v=6$

$C=100\text{km}$

$\eta=20{\cdot}0\text{km}$

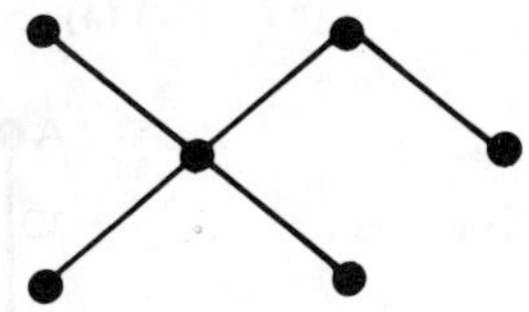

a node = 'ends of edges, intersections and towns of over 50,000 population'

(b) $e=8$

$v=9$

$C=100\text{km}$

$\eta=12{\cdot}5\text{km}$

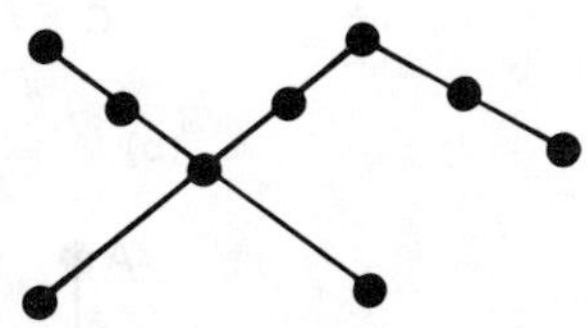

a node = ends of edges, intersections and towns of over 10,000 population'

Weighted measures. Thus far the measures have ignored factors such as distance, node size and traffic flow, but is is possible to incorporate these properties into the analysis. Measures or indices that build in these factors are often referred to as weighted measures, as their value is proportional to the 'weight' of the property.

The Eta Index. The simplest of these is the Eta Index, η, which gives the average edge length:

$$\eta = \frac{C}{e}$$

where:

C = the total distance, in km, of all edges in the network;
e = the number of edges in the network

It expresses the relationship between the network as a whole and its edges as individual elements of the network. It is a useful measure for comparisons between networks, but in such cases it is important that the definition of a node is the same for both or all networks. The more nodes on a network, the more edges it will have between those nodes, and therefore the lower the value of Eta. So, it would be misleading to compare the Eta Index of a network on which nodes are defined as 'ends of edges, intersections and towns of more than 50,000 population' with that of a network on which nodes are defined as 'ends of edges, intersections and towns of 10,000 population'. The same network can yield very different Eta Indices if the definition of a node is altered (Fig. 53.)

A high value of Eta represents a network whose nodes are spaced well apart, possibly because the region is mountainous or unsuitable for economic development for a variety of reasons.

The Pi Index, π. This relates the total distance of the edges, C, in the network, to the distance of all the edges in the diameter of the network, d. In topology the diameter is given by the number of edges in the shortest path between the two topologically most distant nodes. Fig. 54 illustrates the idea of a network's diameter. [δ = the topological diameter (i.e., the number of edges between the two most distant nodes); d = the distance, in km, of the diameter.] So:

$$\pi = \frac{C}{d}$$

where:

C = the total distance of the whole network,
d = the distance of the diameter.

Values are high where the networks are complex, and it has been shown that the index varies in harmony with changes in the level of economic development of the country or region.*

For the networks shown in Fig. 54(a), (b) and (c):

Fig. 54(a)

$$\pi = \frac{65}{65}$$

$$\pi = 1{\cdot}0$$

Fig. 54(b)

$$\pi = \frac{85}{65}$$

$$\pi = 1{\cdot}31$$

* K. J. Kansky, *The Structure of Transportation Networks* (Chicago, 1963), Chapters 2, 3 and 4.

Fig. 54 The diameter of a network

(a)

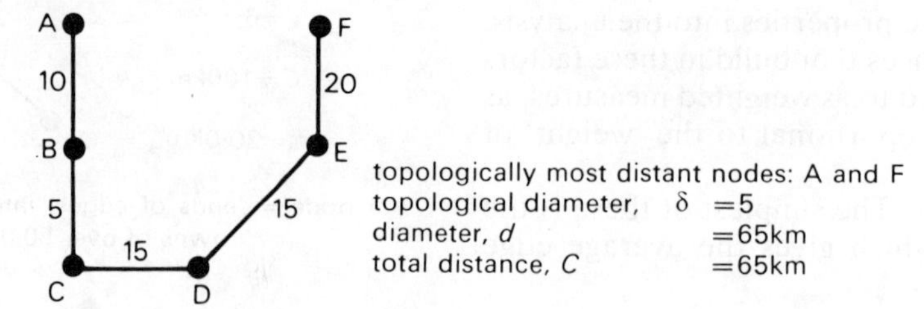

(b)

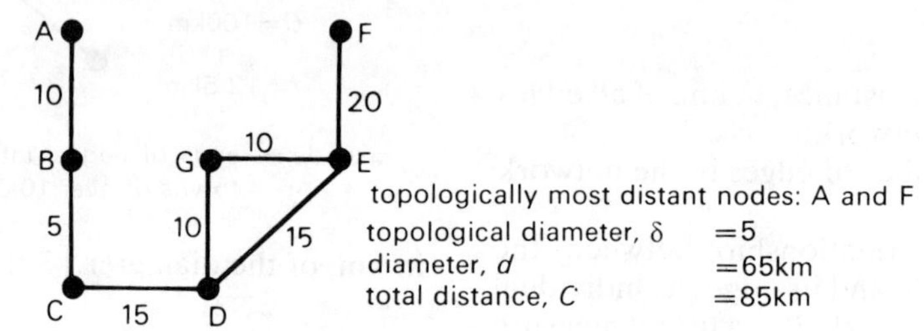

(c)

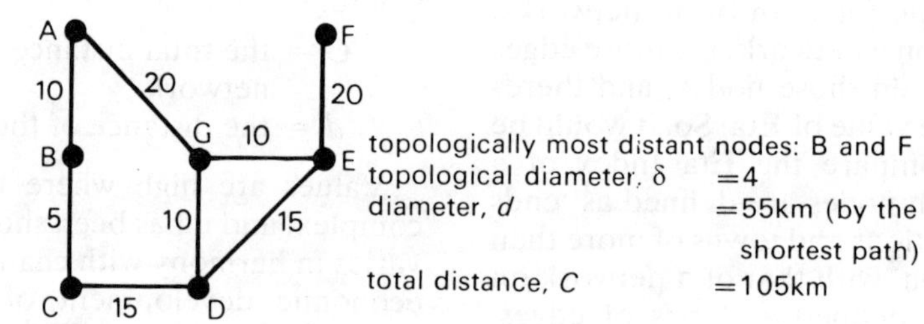

(Distances are given in kilometres next to each edge.)

Fig. 54(c)

$$\pi = \frac{105}{55}$$

$$\pi = 1{\cdot}91$$

Clearly, such a measure may be useful for following the growth of a transport network over a number of years or decades. Comparisons between countries at different levels of economic development may also be enlightening for the student.

There are, however, several limitations to this index. First, the task of measuring the distance of the network is very time-consuming, especially with complex networks. Second, the shape of the country may have an influence upon the result, particularly in the case of long and narrow countries like Chile. Such countries or regions give indices of low value partly because of their long diameter.

Other methods of weighting the indices and of accounting for factors other than topological structure have been devised and used. This is only an introduction to the basis of the technique of network analysis, though, and these methods would be appropriate for more advanced study.

The techniques of transport network analysis rely upon the simplification of the network into a graph and on the counting of the basic elements of that graph. These initial assumptions should always be remembered when results are being interpreted, but the simplicity of the technique is perhaps its major advantage. It enables us to describe and analyse a network in an objective manner and to avoid sweeping generalizations based on nothing but personal whim or the 'look-and-say' method of geographic analysis.

With a knowledge of these techniques and an awareness of their limitations, the quality of descriptions and of geographic analyses will improve.

23 Exercises

1 Fig. 55 shows the motorway network of Belgium.

Fig. 55 The motorway network of Belgium, 1974

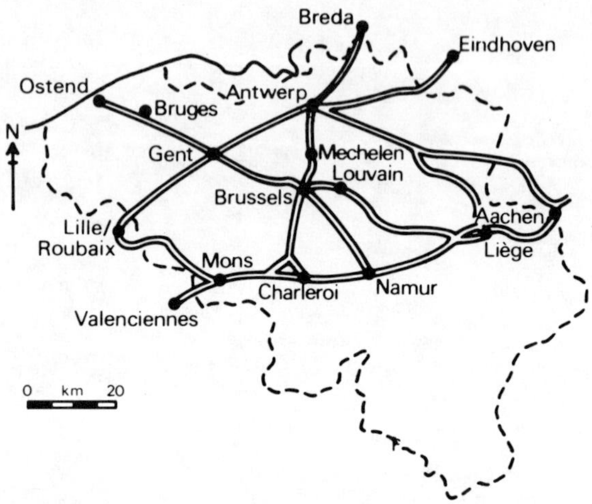

(Nodes defined as ends of edges, intersections and towns of over 20,000 population'. Not all nodes defined as intersections have been marked on the map.)

(a) Remembering that nodes should be included where intersections occur, simplify this network to a graph.
(b) Count the basic elements of this graph—the nodes, edges and subgraphs.
(c) Calculate the Beta Index for this network.
(d) Calculate the cyclomatic number for this network.
(e) Calculate the Alpha Index for this network.

2 (a) Compare the results for the Belgian motorway network with those for the French motorway network shown in Fig. 56.
(b) Try to account for some of the differences between the two networks.
(c) Note which regions in both France and Belgium are poorly served by the motorway networks.

3 The node of Oldham appears on the networks of both the North-West and the Yorkshire and Humberside regions as being relatively inaccessible (Figs. 49 and 51.)
(a) Join the two networks together.
(b) Construct a shortest path matrix for the joint network.
(c) Sum the rows of shortest paths to give the Shimbel Index for all nodes.
(d) Comment upon the relative accessibility of Oldham in the joint network. The different accessibilities for the same node suggest that certain assumptions are made before the technique is applied to any network. What are these?

Fig. 56 The motorway network of France, 1974

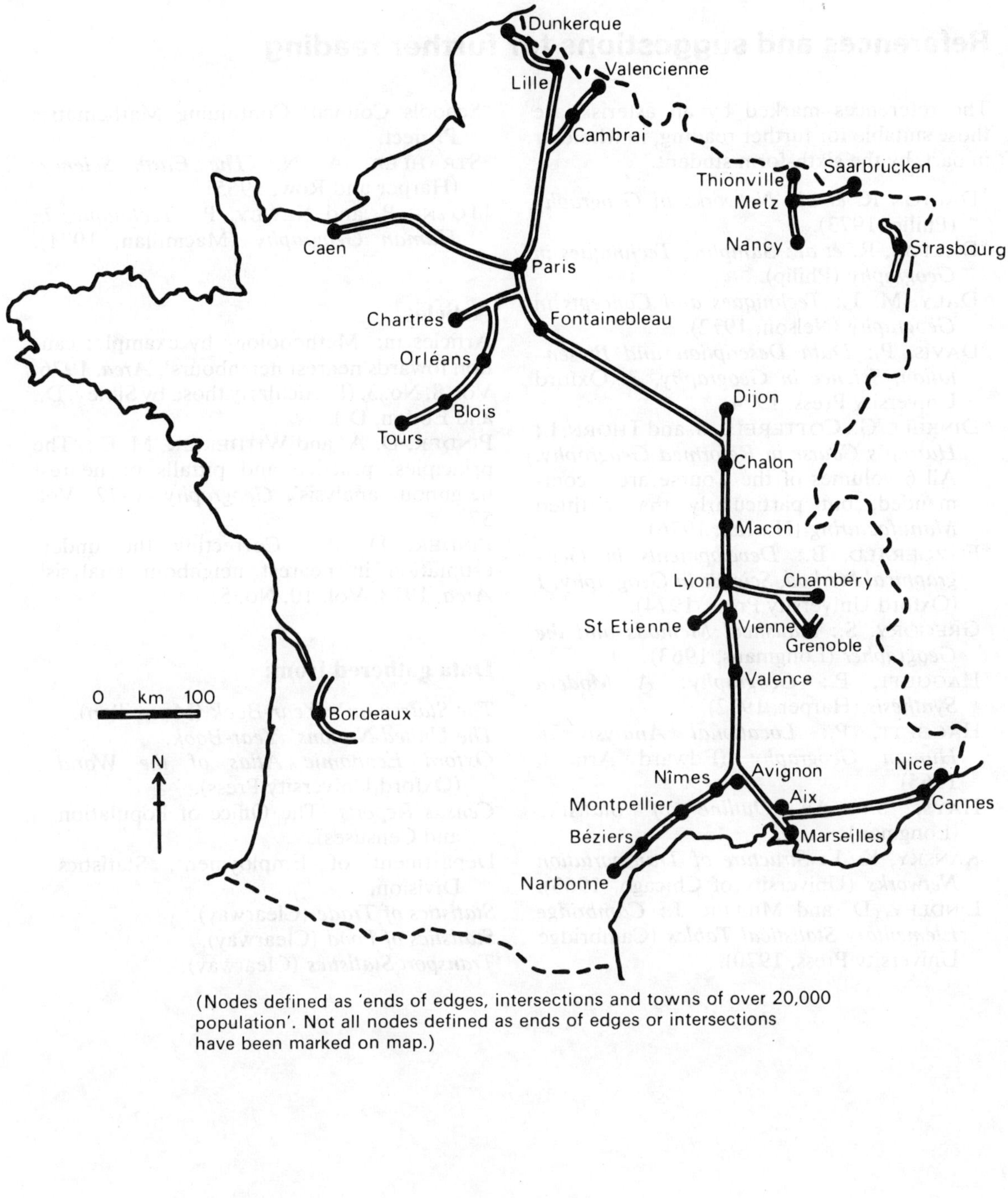

(Nodes defined as 'ends of edges, intersections and towns of over 20,000 population'. Not all nodes defined as ends of edges or intersections have been marked on map.)

References and suggestions for further reading

The references marked by an asterisk are those suitable for further reading, in whole or in part, by the sixth-form student.

*DALTON R. *et al.*: *Networks in Geography* (Philip, 1973).

*DALTON, R. *et al.*: *Sampling Techniques in Geography* (Philip).

DALY, M. T.: *Techniques and Concepts in Geography* (Nelson, 1972).

*DAVIS, P.: *Data Description and Presentation, Science in Geography, 3* (Oxford University Press, 1974).

*DINKELE, G., COTTERELL, S. and THORN, I.: *Harrap's Course in Reformed Geography.* All 6 volumes of the Course are recommended, but particularly that entitled *Manufacturing* (Harrap, 1976).

*FITZGERALD, B.: *Developments in Geographical Method, Science in Geography, 1* (Oxford University Press, 1974).

GREGORY, S.: *Statistical Methods and the Geographer* (Longmans, 1963)

*HAGGETT, P.: *Geography: A Modern Synthesis* (Harper, 1972).

HAGGETT, P.: *Locational Analysis in Human Geography* (Edward Arnold, 1965).

*HAYS, S.: *An Outline of Statistics* (Longman).

KANSKY, K. J.: *Structure of Transportation Networks* (University of Chicago, 1963).

LINDLEY, D. and MILLER, J.: *Cambridge Elementary Statistical Tables* (Cambridge University Press, 1970).

*Schools Council. Continuing Mathematics Project.

STRAHLER, A. N.: *The Earth Sciences* (Harper and Row, 1965).

*TOYNE P. and NEWBY, P.: *Techniques in Human Geography* (Macmillan, 1971).

Articles

Articles in: 'Methodology by example: caution towards nearest neighbours', *Area*, 1976, Vol. 8, No. 3. (Particularly those by Sibley, D., and Ebdon, D.).

PINDER, D. A. and WITHERICK, M. E.: 'The principles, practice and pitfalls of nearest neighbour analysis', *Geography*, 1972, Vol. 57.

PINDER, D. A.: 'Correcting the underestimation in nearest neighbour analysis', *Area*, 1978, Vol. 10, No. 5.

Data gathered from:

The Statesman's Year-Book (Macmillan).

The United Nations' Year-Book.

Oxford Economic Atlas of the World (Oxford University Press).

Census Reports (The Office of Population and Censuses).

Department of Employment, Statistics Division.

Statistics of Trade (Clearway).

Statistics of Food (Clearway).

Transport Statistics (Clearway).

Index

Major references are indicated by bold type